COUNTRY PLUMBING:
Living with a Septic System

COUNTRY PLUMBING

Living with a Septic System

GERRY HARTIGAN

Illustrations by Bob Vogel

Alan C. Hood & Company, Inc.

BRATTLEBORO, VERMONT

Library of Congress Cataloging-in-Publication Data

Hartigan, Gerry.
Country plumbing.

1. Plumbing. I. Title.
TH6124.H27 1986 696′.1 85-30212
ISBN 0-911469-02-8 (pbk.)

Published by Alan C. Hood & Company, Inc.,
Brattleboro, Vermont

10 9 8 7 6 5

ISBN 0-911469-02-8

To Leslie and Hillary, my favorite helpers.

Table of Contents

Preface

Over the years many people have asked me, "How'd you ever get into this business anyway?"

Well, one day shortly after purchasing our house in Vermont we noticed an odor in the front yard. And just like everyone else, I hoped it would go away. It didn't. It got worse. So I called a plumber. Later that day Homer Clark and helper arrived, and soon located the problem.

"Broken pipe."

So, as I watched them fix the pipe, I noted this large container and asked what it was.

"That's yore settic tank," he said in the best Vermont vernacular.

So I peeked in. It was almost full up to the top, where it is supposed to be, but I didn't know that. I assumed it had to be pumped so I asked who did it.

"That fella from Burl'ton," was his answer.

"How much does it cost?" I asked.

"Thirty-five dollars," Homer informed me. I guessed we'd wait a little while.

That was the spring of 1956 and all summer long I noticed the pumper from Burlington in the area. The next time I saw Homer I asked how many people have septic tanks.

"Everybody 'cept them that lives in the village," he said.

His answer convinced me there was a need for this service here. And as I'd had experience with pumps and trucks previously, this seemed to be a natural business for me. So the following spring I bought a second hand tank truck.

One day I went to Burlington to pick up the pump for the tank, and as I drove back through the village, there stood Homer.

He waved me over.

"That yore rig?" he asked.

"Yes."

"When you gonna git it going?"

"Should have it ready in a couple of days."

"Got a job for ya."

I was delighted. I hadn't even started and the work was coming in. "Whose?" I asked.

The answer, I'll never forget. "Mine, goddammit."

And that's how it all started.

mine damn-it!

COUNTRY PLUMBING:
Living with a Septic System

CHAPTER 1

The Intent

After twenty-five years of installing and servicing septic tanks, I'm still amazed to find out how little people actually know about their home sewage disposal systems. And this is true, not only of people who have recently moved to rural areas, but also of those who have lived with their different systems for many years. As Bismarck said, "The less people know about politics and making sausages, the better they sleep at night." Perhaps this is also true of septic tanks.

However, with increasing numbers of people moving from urban to rural areas, or buying second homes there, it's more important than ever to understand more about what is generally referred to as "country plumbing."

It is, therefore, the intent of this book to help familiarize these people with the different types of systems now in general use, and the problems which accompany them.

alice the plunger!

CHAPTER 2

The Universal Problem

You bought your house seven years ago and it was five years old at the time. Since then everything has seemed to function pretty well. Once you had to call the electrician to fix the oil burner, and once the TV man to fix the set. Other than that, no big problems.

This morning seems just like any other morning. You've just finished your morning constitutional and it happens. The water in the toilet bowl starts its ascent to the rim. Damn! Well, the manufacturers seem to design these units so the water just reaches the top of the bowl before it stops.

The next move is to scream, "Alice, get the plunger. The toilet's plugged up." A few minutes later the plunger arrives, a little dusty, but still serviceable.

You address the bowl with this instrument and proceed to try and stuff the flexible head down the toilet's throat. After several frustrating and unsuccessful attempts (and a slight displacement of water), you retrieve the plunger and notice that the little head has somehow turned itself inside out. And now you are looking at a very small orange-colored umbrella.

How to pop it back into its original position? Simple. Just place it between your feet and yank up. You almost have it and "plop"—your nice clean pants are now ready for the cleaners. At this point the language deteriorates rapidly.

"Alice, call the plumber."

For a change, something goes right and the plumber says he'll be there right after lunch.

Sure enough, he arrives around two o'clock, in the usual beat-up van, with helper in tow. Everything's going to be all right now. With much self-confidence and one of those new scientifically designed plungers—you know, the ones you see advertised in magazines—he proceeds directly to the problem area.

Prior to his arrival, you have meticulously mopped the floor, as you didn't really want him to know you'd failed with this seemingly basic tool. However, there's a secret to using this instrument. Instead of trying to push the stoppage down the pipe, the plumber slowly pushes the plunger down and then quickly yanks it back up. If the plunger is going to work at all, this is how to use it. In this case, it doesn't.

So the plumber removes the plunger, unceremoniously plops it on the clean floor, turns to the observing helper, and says, "Donald, git the auger."

Several minutes later Donald returns carrying this tube with a crank handle on its top, and you're informed, "This is a toilet auger." And if you wondered why it took him so long to find it, later when you've had a chance to view the inside of the van you're surprised he found it at all, as the inside of the van looked as if it had been turned upside down at some time or other.

Once again, with professional aplomb, the plumber draws about three feet of flexible wire into the tube, and inserts the end with a small bulge on it into the toilet. Now, by pushing and cranking at the same time, he forces the cable down into the toilet. This simple operation is repeated a half dozen or so times. No luck. At which time he turns and solemnly announces, "Yup, it's plugged up, awright."

It's always reassuring to have someone tell you what you already know. Now what?

"I guess it's your septic tank," he mutters. "You know where it is?"

"No," you murmur weakly. "I didn't know we had one."

"Know who built the house?" he asks.

"No, and the man we bought it from died."

"That figures," he sighs. "Let's go into the basement and have a look."

Well, there it is, the big black pipe going through the cellar wall. The location of this pipe tells you two things: which side of the house the tank *may* be on, and also how deep the tank *may* be. Why do I say "may" be? Because once outside the wall, the tank may be anywhere. Also it may be at any depth. Installers do strange things. Many times, regrading after the house is built puts the tank much deeper than when it was originally installed.

But at least we have a starting point. If the pipe is more than three feet below grade outside, I recommend hiring a backhoe to locate the tank.

If it's less than that and the soil is not too hard, there's a good chance you can find the tank with a long probe rod. Now the treasure hunt begins, and it's the helper's turn to go to work. Under the supervision of the plumber, who has a large chew of tobacco in his mouth as well as a shovel to lean on, the helper starts to aerate your front yard.

Now sometimes it's possible to find the tank by removing the cleanout plug at the end of the big black pipe in the basement just before it goes through the wall. Insert a sewer rod and by pushing the rod out, you should hit a hard stoppage that should be the inlet baffle of the tank.

Now two things can help. By measuring the rod, you can get a general idea of how far out the tank is. Also, by pushing the rod in and out and rapping the baffle, you can sometimes locate the tank by having someone listening outside. I've found many tanks this way.

However, there's a slight problem involved with this procedure, and that's taking the plug out in the basement. *That* can cause disaster: a flooded basement and a very unhappy homeowner.

If you do attempt to remove this plug, do so very carefully and slowly, as there's almost always pressure behind it. If the tank is plugged, water will start to squirt out as soon as you start to loosen the plug, but you'll be able to tighten it back in before it flies off and floods everything.

I'm not concerned with the prospect of the homeowner removing this plug, as they are usually, it seems, put in originally with a four-foot wrench. Only a plumber with powerful arms and vocabulary is going to get it out.

How we find the tank—by rod, backhoe, or just plain luck—is unimportant. But let's say we find it. The first thing to do is remove the cover and look in. This will tell the plumber or septic tank man all he needs to know. In this particular case, the twelve-year-old tank is long overdue and pumping the tank eliminates the problem.

However, there is one small consolation. Save that little orange plunger, the next time your neighbor, the one who borrows everything and returns nothing, needs a plunger, let him take this one. He'll never borrow anything from you again.

Since we're on the subject of "plumber's helpers", also called plungers, you may wonder sometimes why they perform so poorly. The answer is simple. You are probably using the wrong one, not using it correctly, or both.

There are two types of plunger. The toilet plunger and the sink and drain plunger. (See Figures 1 and 2.) The toilet plunger is designed to work in the toilet bowl only, while the sink plunger is for kitchen sinks,

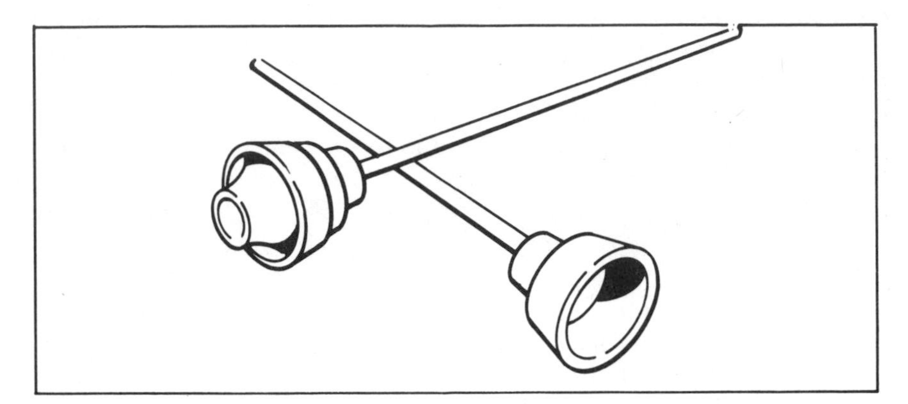

Figure 1. The Toilet Plunger   Figure 2. The Sink And Drain Plunger

bathtubs and drains with flat surfaces. The toilet plunger will not work on a sink drain, and if you try to use the sink plunger in a toilet bowl, as millions have, the results are poor.

As I stated earlier, the proper procedure is to push down slowly, then yank up quickly, as if you are trying to pull the stoppage back out.

If you have a plugged lavatory or tub drain, remember to hold a damp rag over the overflow holes while you're using the plunger. Otherwise you'll be there all day.

My advice is to have both plungers available. If you have a commercial establishment (motel, restaurant, gas station, etc.) by all means invest in a toilet auger (Figure 3), and a sewer rod (Figure 4). They could save you a lot of money. These tools are available in most hardware stores and at any plumbing supply house.

What you have read so far happens thousands of times every day. It's a universal problem appreciated only by the plumber and the septic tank service man.

Now we'll go into the subject further, and explain in more detail the workings of the septic tank and other sub-surface disposal systems as well as the problems associated with them.

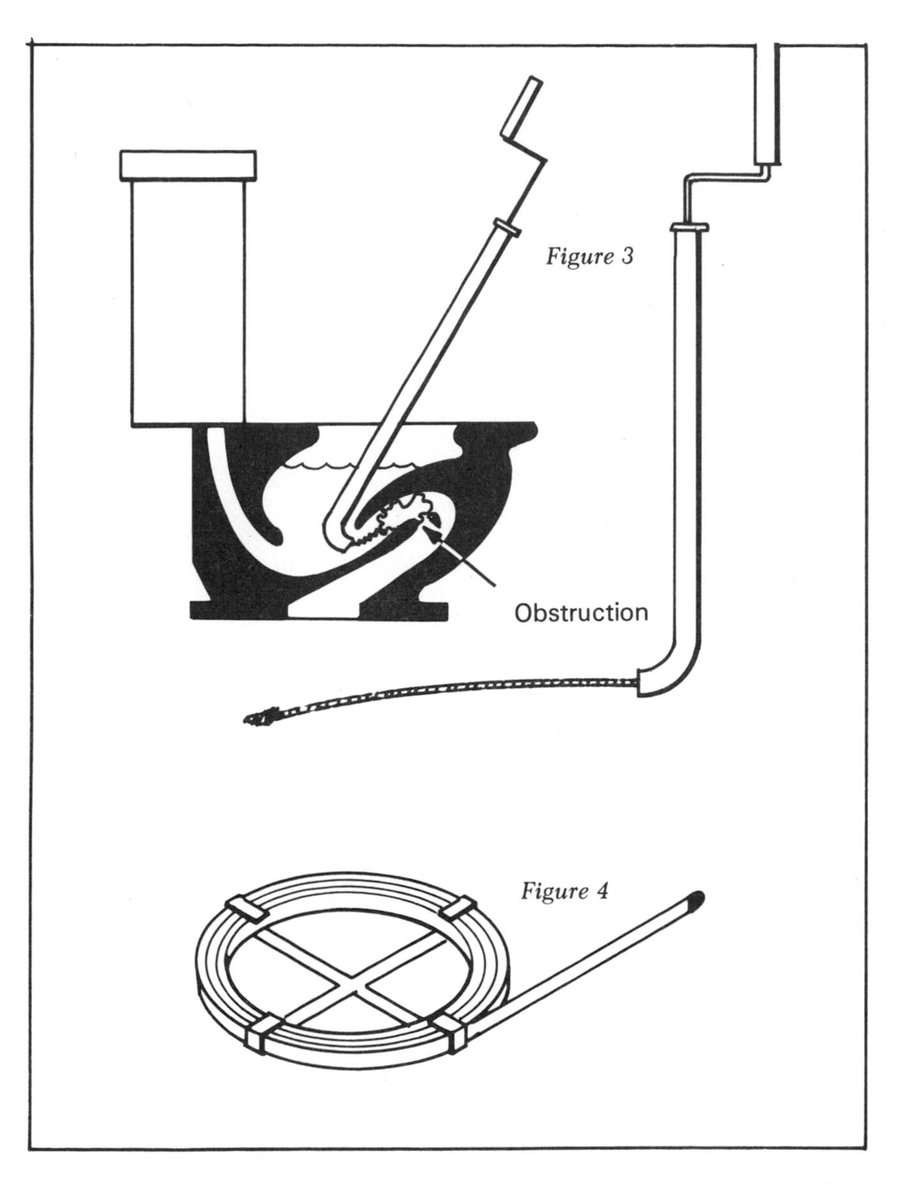

Figure 3. The Toilet Auger

Figure 4. The Sewer Rod

CHAPTER 3

Background

Let's begin with the definition of "sewage." Very simply, sewage is liquid waste from residential and commercial buildings. Since we're concerned only with rural, residential, and commercial systems, we'll cover the four basic disposal systems in order, from the earliest and simplest to the more complete modern systems. They are the privy, the straight pipe, the cesspool and finally, the septic tank.

In early rural America, before running water, people had to carry water from either a brook, a river or a well. In certain areas, the spring was the prime supplier of water. Water was delivered to the house through wood, lead or galvanized pipes. Plastic pipe was still a long way in the future. No matter what the source of water, it was always in short supply. And this brings us to the need for "outdoor plumbing", more commonly known as the privy.

The Privy

The privy was pretty much universal in design, a shack on top of a hole in the ground. The original time-share unit. Now, this system represents one of the truly uncomplicated sewage disposal systems man has ever known, requiring no energy, no special training to use, and virtually no maintenance other than moving the building over to a new hole occasionally. Like all good sewer systems, in order to function it depended only on Newton's Law. Gravity.

Basically, privies were made pretty much the same, much like Model T Fords. The number of seats would vary according to traffic.

The location, in many cases, was determined by the easiest digging site and how quick-a-foot Grandpa was. The privy, also known by other names, was no place to socialize, particularly in northern regions. I cannot imagine all that sensitive equipment being exposed to the elements for too long. But Grandpa had the solution to that problem. It was handy. It had a handle. It was under the bed. And Grandma called it the "chamberpot."

The privy is discussed because even today it has its place and use. Not only is it environmentally acceptable, it's functional, and in many locations, like deer camps and backwood sites, it's the only system available. And for all its shortcomings, a privy is more pleasant to use than the so-called chemical toilets. Even in winter.

If you live in the wide open spaces and decide to build a privy to serve your needs, here are a few things to consider.

First, pick a site that's at least 100 feet from your water supply, preferably downhill from it, and downwind from the house.

The number of people it serves determines the size of the hole you must dig. Four feet deep is about average. The bigger the hole, the longer it can be used before relocation. A hole three feet by four feet and four feet deep will serve a family of four for many years.

When wastes pile to within eighteen inches of the top of the hole, it's time to move the privy. Fill the old hole with earth, dig a new one, and set the building over it.

Old-timers used to dig a hole, then line it with a wooden crib to keep the sides from falling in. They would mount the privy on top of the crib. Sometimes the privy was attached to the house, so it could be reached without going outside. This was convenient, but meant the accumulated wastes had to be removed occasionally.

If you're planning a permanent location, try lining the hole with a circular wall of concrete blocks. They won't rot out, like the wood will. The building can be set on top of this wall.

The design of the building is pretty standard, as I said earlier. Basically it is three sides, a roof, and a door. Inside is a bench arrangement with a hole that has some sort of lid covering it. In some cases a toilet seat is used. Nothing like a little touch of class.

If the privy is to have a lot of traffic, attach a large plastic funnel to the wall, with a short plastic pipe running into the main hole. This will help tall soldiers with short rifles to score better, and improve the area around the seat considerably.

There are problems associated with the use of the privy. The first is accompanying odors. These can be controlled in various ways. A scattering of chlorinated lime, available at most hardware stores, applied after each use, helps a lot. Dirt is sometimes used, but it will fill up the hole faster than lime.

The second problem is the ever-present housefly. Screen all openings in the building, keep the door shut at all times, and hang up flypaper for those flies that get past these barriers. Using a thumbtack and hanging the flypaper inside, under the seat where the most fly activity takes place, seems to work best.

If you live in a place where hedgehogs abound, be prepared to replace the seat from time to time, as these little fellows will manage to destroy it. A sliver from this area can be extremely uncomfortable.

midnite caller

The Straight Pipe

The advent of indoor plumbing occurred in different areas at different times. However, with it came the typical problem of what to do with the waste water. Granted, the earliest indoor plumbing was a far cry from what we have today, but it was a big improvement over the privy, by anyone's standards. Again, simplicity and gravity came to the aid of rural America.

Where one lived determined the direction of the sewer pipe. As most of the population in rural towns lived near a river or stream, and since they ran through the lowest points in the valley, the natural direction was a straight line to the water. In many small towns during the 1880's a sewer district was formed, and the people of the district were allowed to join with their neighbors in a communal "straight pipe" system. This system was usually installed by the town or village and the users were taxed accordingly.

But what about the people outside the sewer district? They could always run the pipe to the nearest available place, be it a brook, ditch, or open field. (But this led to health and other types of problems and better systems were soon to follow.) In any event, this was a trouble-free and inexpensive system to operate.

Figure 6. The Straight Pipe System

The Cesspool

We now come to the cesspool, and once again the simplicity of this system is hard to believe. It's merely a hole in the ground lined with stone, brick, or cement blocks into which sewage flows. No design criteria — just a hole of any size or shape, though usually round and shaped similar to a beehive.

Occasionally, after spring thaws or heavy rains, cesspools overflowed. But that presented no big problem once the water table stabilized and things returned to normal. Once in a while cesspools required pumping, and every now and then a cover would fall in. No problem. Wood was cheap and soon a new cover would replace the old one. Sometimes a sewer pipe would be run into an old abandoned well, and this saved the trouble of having to dig or build a new one. Nothing is wasted in the country.

Personally, I feel a large cesspool in *proper soils* is a good disposal system. It is an efficient and inexpensive sub-surface disposal system. But be sure to check local regulations before installing a cesspool.

Figure 7. A Typical Cesspool

The Septic Tank

This is one of the few things in the world that the Russians do not claim to have invented. The septic tank is supposed to have originated in France and developed later in England. We'll describe the most popular ones in use today. They're generally made of steel or precast concrete and come in a wide variety of shapes and sizes. Let's begin with steel tanks. There are basically two types: vertical and horizontal. Below is a diagram of a typical vertical tank.

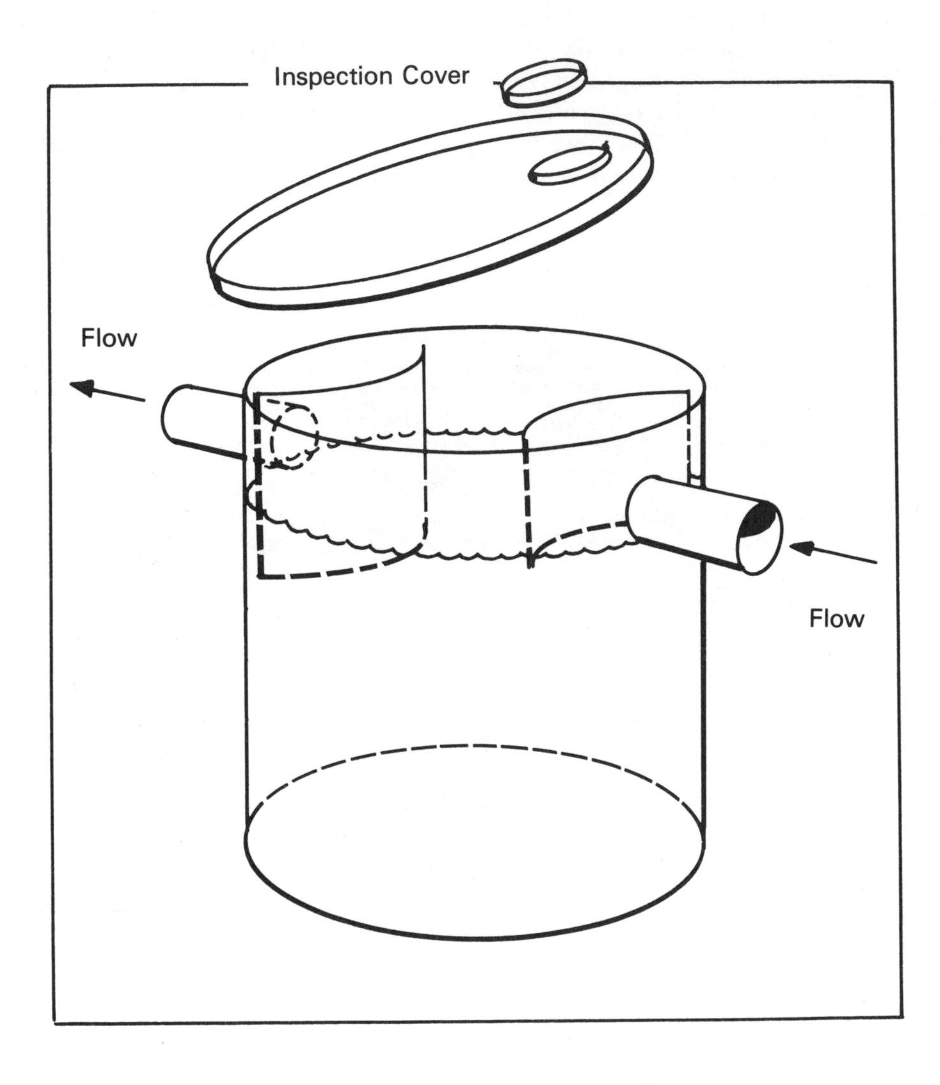

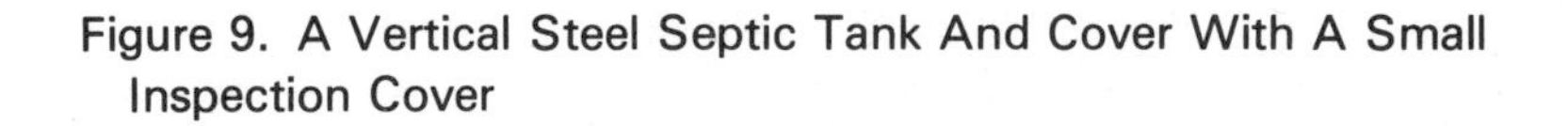
Figure 9. A Vertical Steel Septic Tank And Cover With A Small Inspection Cover

It resembles a coffee can with a removable lid. The better quality tanks have nuts and washers to secure the covers. Less expensive ones have metal clips, and all are coated with tar for corrosion protection.

Some vertical tanks will have a small cover over the inlet baffle. This is for inspection purposes only, but if you have a simple stoppage at the inlet baffle, it's handy to remove this small cover and poke the obstruction free. Also, if there's a stoppage between the tank and the house, this is the place to rod it from. Unfortunately it's virtually impossible to pump a tank properly from this small opening as the inlet baffle is directly below it making access to the main section impossible. In this situation the entire cover should be removed to pump the tank.

Shown here is a bird's eye view of a vertical tank, with the cover removed, showing the baffle positions. (Figure 9)

You should now see why it's not recommended to pump through the small cover, as the hose can only go inside the inlet baffle and to the bottom, removing only the liquid wastes. And leaving the solids.

The following description covers horizontal steel tanks:

The inside layout of the horizontal steel tank is the same as the vertical tank, except that a 1,000-gallon horizontal tank will sometimes have a full baffle three-quarters of the way down inside the tank, just in front of the outlet baffle (See Figure 10).

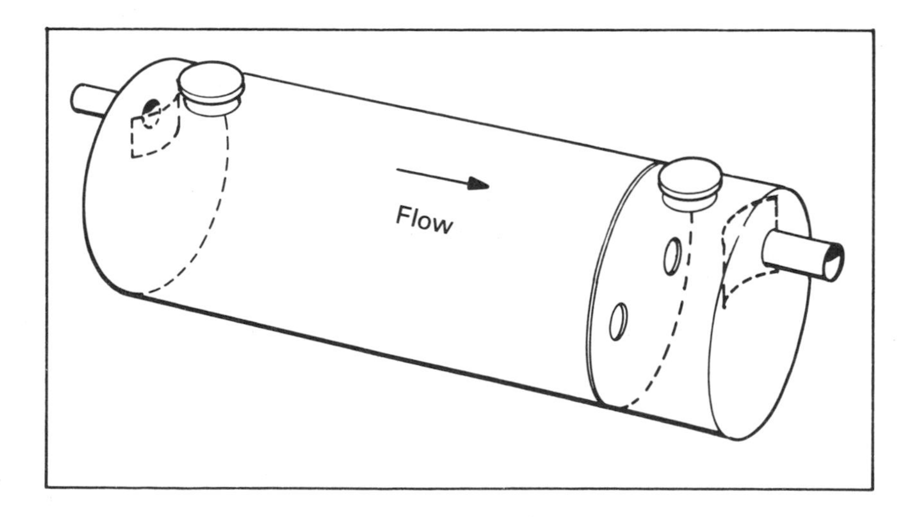

Figure 10. A Horizontal Steel Septic Tank

Horizontal steel tanks should always be pumped from the inlet cover. If the full baffle is present, it may be necessary to open the-outlet and empty the tank completely.

Horizontal tanks seem to perform much better than vertical ones. The reason for this is a simple design factor. The length of the digestion area is generally twice the width and that's just what it's supposed to be.

Listed below are some sizes and capacities of vertical and horizontal steel tanks. There are larger tanks available, of course, but since we're concerned only with residential and small commercial sizes, these are not listed. Dimensions are given here so that if you have to dig up your tank, you'll have a pretty good idea of its capacity.

Vertical:	Diameter	Depth	Capacity
	4′	4′	300
	5′	5′	500
	5′	6′	750
	6′	7′	1000

Horizontal:	Length	Diameter	Capacity
	5′	5′	500
	6′	5′	750
	10′	4′	1000
	15′	5′	1500

By now you have figured out that a horizontal tank is basically a vertical tank installed on its side, the difference being only the inlet, outlet holes, and cover arrangement.

Finally, we have the precast concrete tank. These tanks are generally cast in two sections, a top and bottom. Both halves are delivered to the site by the manufacturer or installer and there set in place. The top and bottom are then sealed together with a mastic compound to make the tank watertight. Today the concrete tank is the most widely used.

By now you also realize that wastes come in one end of the concrete tank and leave by the opposite end, the same as in the steel tanks. There's a slight difference, however, in the cover arrangement. The inlet has an inspection and clean-out cover only, and should a blockage occur here (or should rodding to the house be necessary), this is the place from which to work.

The center cover, which may also be round in shape, is removed for pumping. The little cover on the far end, the outlet cover, is very seldom

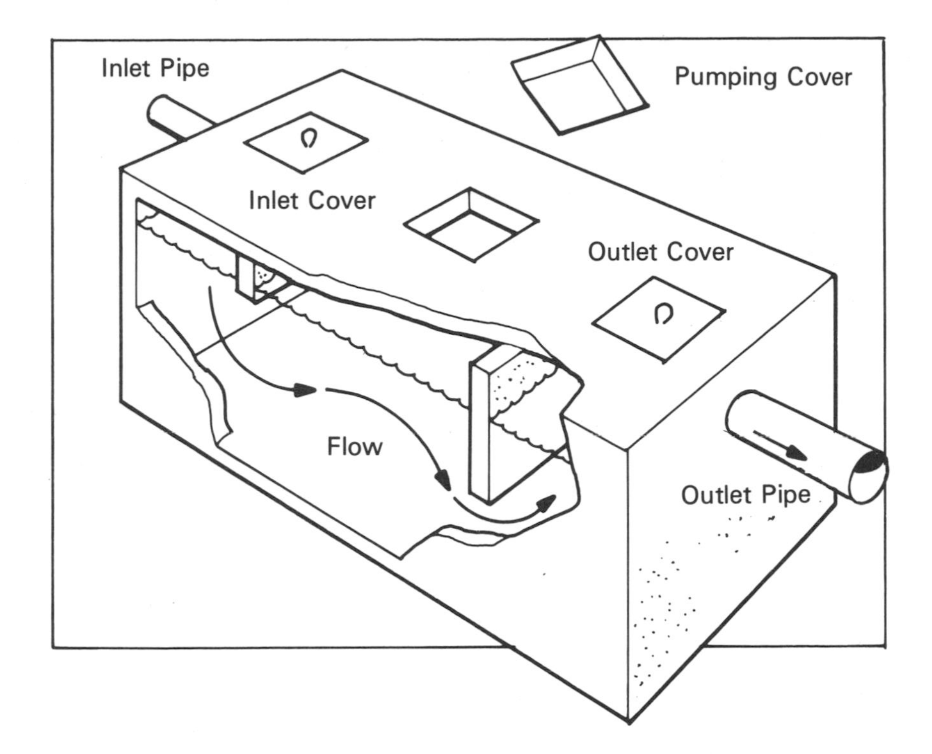

Figure 11. A Cross Section Of A Typical Concrete Septic Tank

opened, except when there's a plugged baffle which needs rodding or a stoppage between the tank and seepage system.

Now, the alternate inlet is another story. This four-inch knockout section allows the tank to be installed at right angles to the house. The sewer pipe can enter here after the thin shell of concrete is punched out. There's another knock-out just like it on the opposite side, and of course, there's the regular inlet at the front end of the tank. (Figure 12)

Figure 13 shows the inlet and outlet serving a single entry and exit.

Figure 14 shows a multiple access tank. If the tank, for one reason or another, *has* to be installed at right angles to the house, the sewer pipe can enter on either side. And this is where one problem begins.

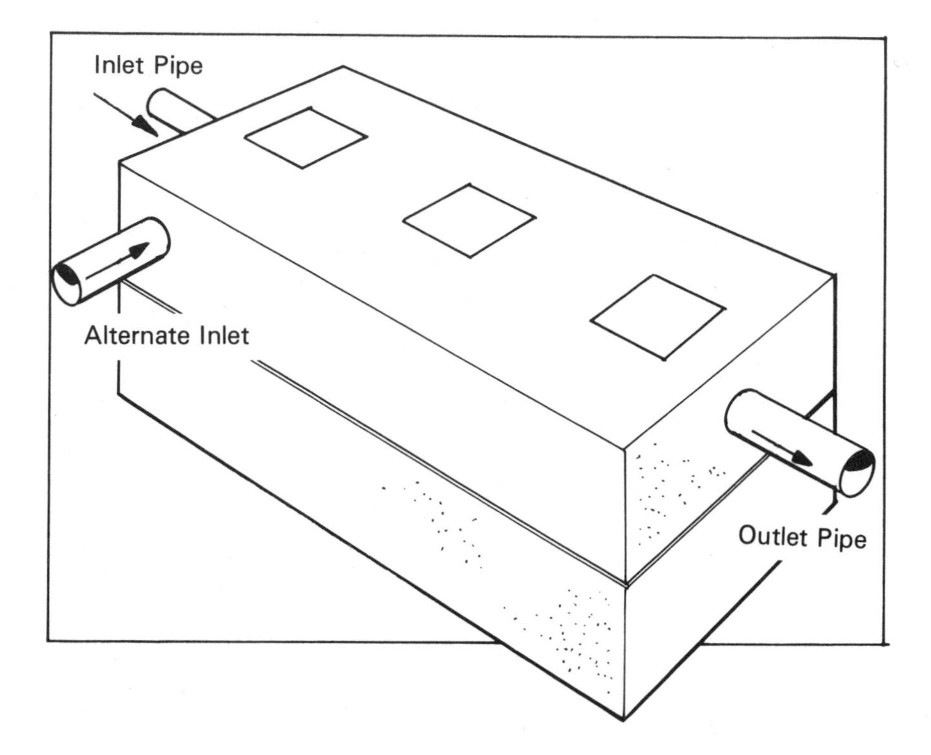

Figure 12. A Concrete Septic Tank With Knockout For Right Angle Installation

In the first illustration the inlet baffle encloses a small section (this is good), whereas in Figure 14 the baffle resembles a full partition that extends across the tank at about one-third of the total liquid depth (this is bad). The problem here is that solids arriving at the front of the tank float around leisurely, don't get into the digestion area quickly enough, and begin to build up in this small section. This can cause a continuing problem.

If the tank is to be installed this way the inlet line should have two ⅛ bends to encourage better flow. Quarter or 90 degree bends should never

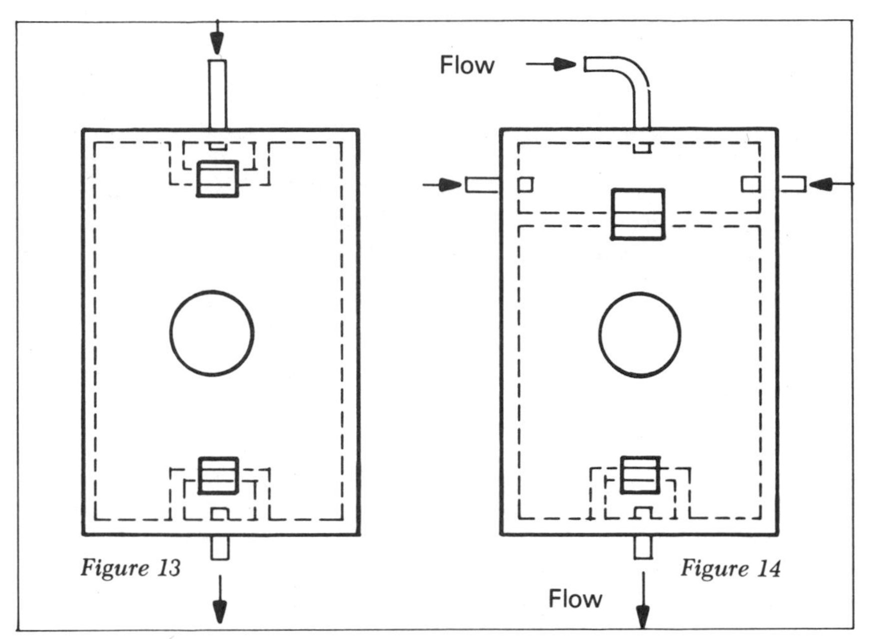

Figure 13. Top View Of Single Entry And Outlet Septic Tank
Figure 14. Top View Of Multiple Entry And Single Outlet Septic Tank

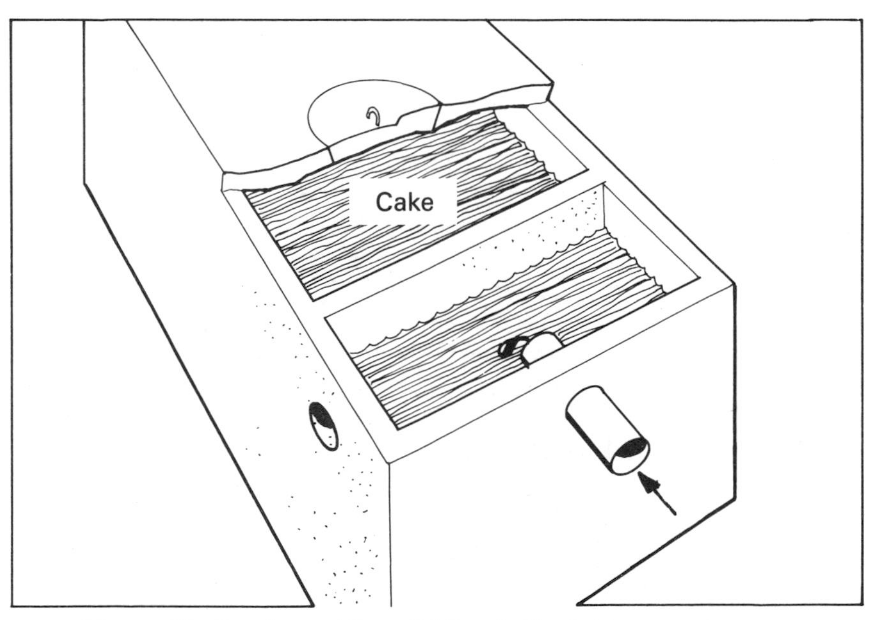

Figure 15. Cut-Away View Of Septic Tank Showing Cake-Up

be used between the house and the tank, for two reasons. First, should the sewer line have to be rodded out, the sharp bends make it very difficult to work the rod effectively. Second, they reduce flow and hinder the entry of solids into the tank and digestion area. At the outlet of the tank, on the other hand, any type of bend may be used.

Next, any sewer line that has to be rodded from a tank with a side entry is very difficult because it's so hard to get a sewer rod into the opening. It's always better to open a sewer line from the tank end if possible. This allows the plumber to be sure both sewer line and baffle are clear. Further, any water pressure built up in the line helps the obstruction into the tank, not all over the cellar floor.

As you can see from the illustrations, sewage enters at one end of a concrete tank, and effluent runs out the opposite end, just as in steel tanks.

But there are differences. The inlet cover is for inspecting. If there's a blockage in front of this inlet, or if it's necessary to rod to the house (that means poking a rod into a pipe to free an obstruction), this is the place from which to work.

Remove the pumping cover when the tank is to be pumped. The little outlet cover is rarely opened, except in cases where there's a plugged baffle or a stoppage between the tank and seepage system that requires rodding.

The alternate inlet is another story. There's one on either side of the tank, placed there so that one of them can be used if the tank must be placed at right angles to the house. (Figure 12)

My advice: don't use them. Solids will build up in this small area instead of the larger digestion area, causing unnecessary back-ups, and making it more difficult to rod out an obstruction in the pipe leading into the tank.

If you must install the tank at right angles to the house, avoid using an alternate inlet by placing the tank in such a way that two ⅛ bends will lead the pipe to the front end of the tank.

If you *have* to use the side entry hole, the best way is to insert the pipe into the tank with a "sanitary tee," as shown in Figure 16. A short extension reaches from the tee down into the tank. This extension pipe should reach into the tank so the tee is directly under the inlet cover. The outlet pipe can go in any direction.

If you don't require this side-hole arrangement, and have your choice of a tank with or without these side-holes, pick the one without them. The baffle systems in the two tanks are different, and tanks equipped with the regular baffle arrangement are far more satisfactory.

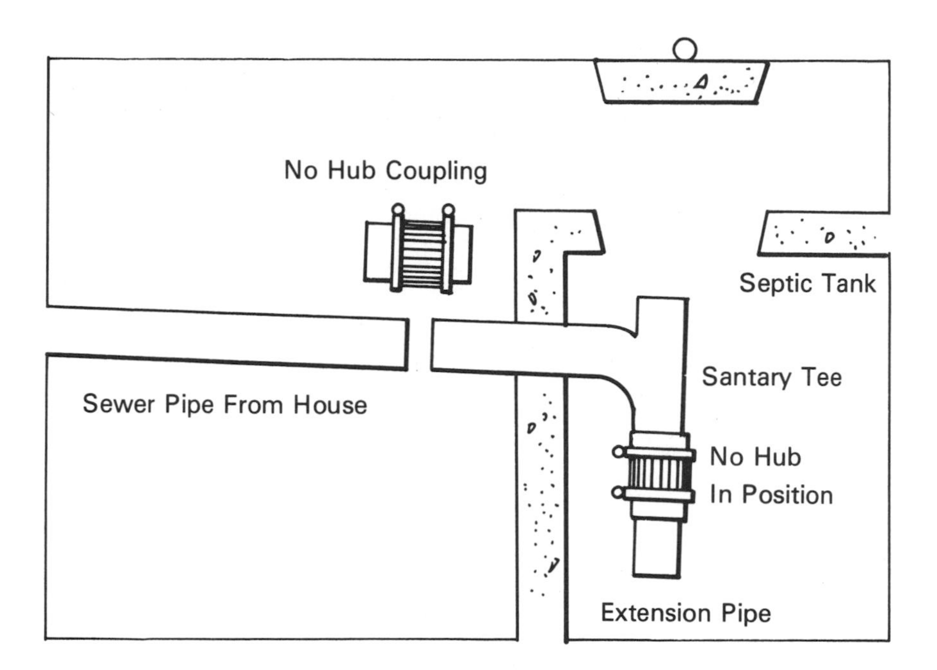

Figure 16. Sanitary Tee Assembly Installed For Side Entry Septic Tank

Note the difference in baffle arrangement between the tank with a single entry and the one that provides a choice of three entries. The first tank operates with fewer difficulties.

Concrete Tank Capacities and Sizes

Capacity gallons	Length	Width	Depth
750	8′1″	4′	5′4″
1000	8′6″	4′10″	5′4″
1250	12′	6′6″	3′11″
1500	12′	6′6″	4′5″

How The Tank Works

The accompanying illustration may be the most important one in the book. It shows how a septic tank operates, and all septic tanks work this way, no matter what their size, shape, or type.

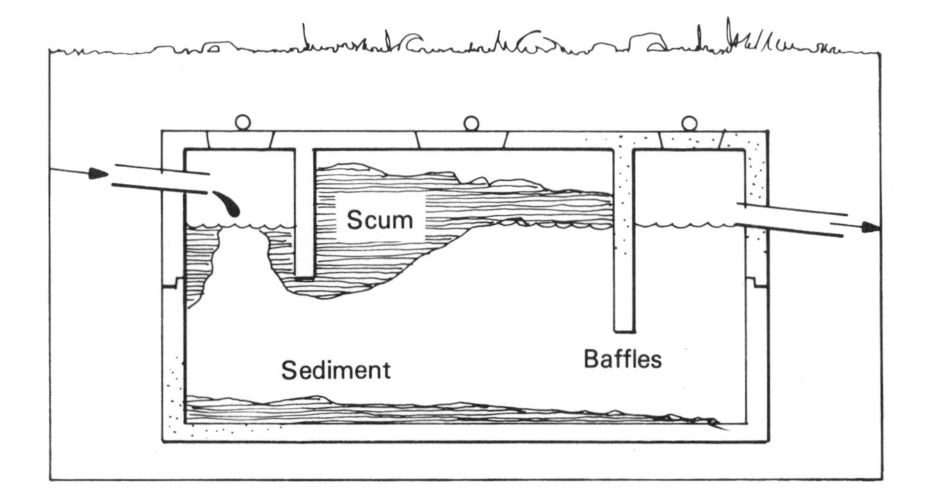

Figure 17. Section Thru Septic Tank Showing Result Of Neglect

Between the two baffles is the digestion area. The dark mass is referred to in many books as the *scum*. I call it the *cake*, because I think it's more dignified, and since I've always called it that, I'm not going to change now. In this area the floating cake is broken down by bacteria into gases and water. The residue settles to the bottom as the *sediment*.

Contrary to most reports, the cake is not of uniform thickness, and as the diagram shows, the thickest section is usually just inside the inlet baffle. It is generally three to four times thicker here than at the opposite end. Also, the amount of sediment at the bottom of the tank is greatly exaggerated. You'll notice that the accumulation of such garbage disposal wastes as egg shells, coffee grounds, etc. are pretty insignificant. More on the mechanical pig later. In the meantime, the owner of this tank will be calling the plumber in the very near future, as I'll explain.

What does all this mean? When waste matter leaves the house and travels to the tank, it contains floating and suspended solids. It enters the tank, drops under the inlet baffle, then floats into the digestion area between the two baffles. In this area it's acted upon by all kinds of bacteria, the names of which are unimportant, and here the primary digestion process begins. The effluent leaves the tank and travels to a disposal system of some kind, a series of pipes, usually, which spreads it through the soil.

When the septic tank is in operation, sewage flows in and down below the first baffle. There it begins to separate, with a dark mass called the cake floating on the surface, sediment settling to the bottom, and a liquid, filling most of the tank. This liquid gradually spills through the outlet beyond the second baffle.

The difference between a cesspool and a septic tank should now be obvious. The cesspool receives sewage and digestion, settling, and leaching all take place in one area.

The septic tank on the other hand receives, digests, and partly clarifies the sewage in the tank, before the effluent is absorbed and treated in another area.

There are only three reasons a septic tank or disposal system can fail: poor design, poor installation, and finally and most common, poor maintenance.

In Figure 17 you see a large accumulation of cake. How does this happen? Neglect. The material that arrives in this area is supposed to remain here and be digested for between twelve and twenty-four hours. Sometimes the solids arrive faster than the digestion area can handle them, or the tank and its digestion area are simply too small. Sometimes tanks are without sufficient bacteria to function properly. Consequently, the cake starts to build up, and as this mat floats, it dries out. When this happens, the digestion process stops and the tank becomes a holding tank for solids. Also, as you can see, the normal air space is greatly reduced, causing waste water from the house to have difficulty entering the tank.

The first indication of this will be a gurgling sound in the house—at the kitchen sink or bathroom drains. The toilet flush is slow. So is the washing machine waste cycle.

Trouble is just around the corner.

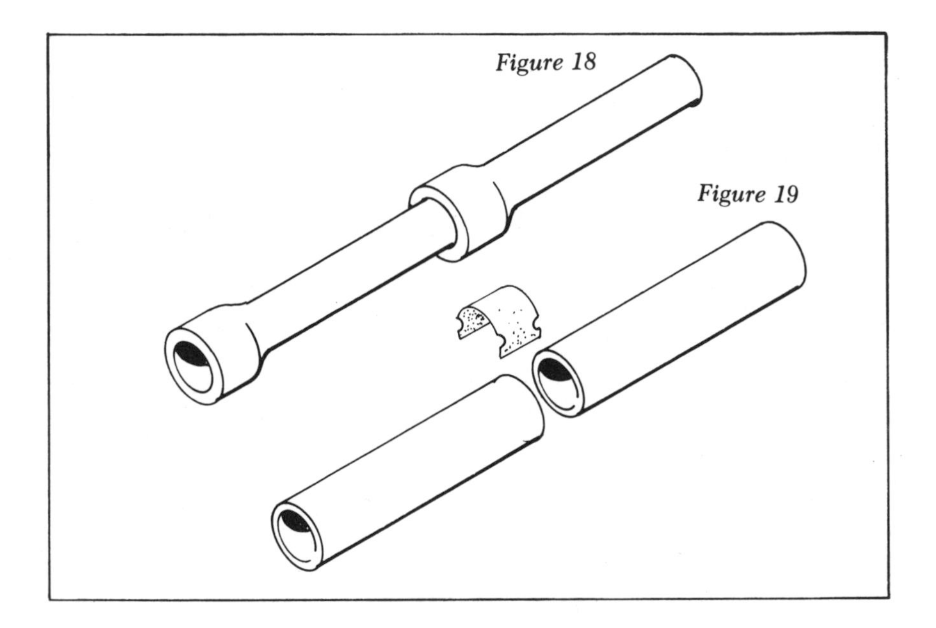

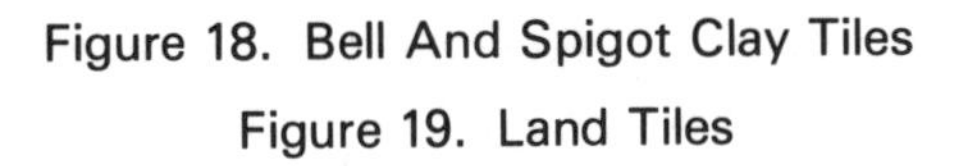

Figure 18. Bell And Spigot Clay Tiles

Figure 19. Land Tiles

CHAPTER 4

Subsurface Disposal Systems

There are four types of disposal systems in general use today. They are leach fields, leach beds, seepage pits, and dry wells. There are, of course, other systems, but these are the most widely used and when installed and maintained properly, they should be trouble-free. Figure 18 shows two sections of bell and spigot clay tile. When installed, the small end of one section is shoved into the large end of another section and they're cemented together. These sections are four inches in diameter and two feet long. With them come the usual assortment of fittings: Y's, T's and bends. These were the solid pipes of not too long ago.

For drainage, another type of clay pipe was used. These were one foot in length and again, four inches in diameter, sometimes called "land tile." (Figure 19).

The funny little object on the top is a piece of tarpaper used to cover the tops of the joints. The little ears were tucked inside to hold them in place to keep out the dirt, and anything else.

These pipes were used to allow effluent from the septic tank to seep into the ground. They were used also to drain lowlands or wet areas and were very successful in both applications. Once in a while the solid pipes, if not cemented properly, would allow a tree root or two to get inside. Also, they could not take too much abuse and needed to be protected. Other than that, they worked very well.

These pipes were also available in larger sizes for municipal services, but four inches was the most widely used for residential systems.

The next type of pipe to arrive on the scene was known as Orangeburg, Bermco, and various other trade names. This was an impregnated fiber tube which came in three lengths: five, eight and ten feet, and two models: solid and perforated. The solid was supplied with solid tapered couplings, and the perforated came with split couplings. The usual assortment of fittings was also available. They came in one color. Black.

The next material to replace Orangeburg and cast iron was Transite, a composite of asbestos and cement. This was generally used outside the house where solid pipe was needed.

I have not forgotten cast iron, but since this is universally used for sewer pipes inside the house and goes through the cellar wall, I don't think it's necessary to describe it here.

The most recent breakthrough in pipe is called Polyvinylchloride, or PVC for short. These pipes come in ten foot lengths in both solid and perforated models. They are white and very light in weight, some with solid couplings already moulded on one end.

As with so much else in life, new is not always better. In previous illustrations you saw how the joints in both clay tile and perforated Orangeburg are not tight, but open slightly. This allowed the pipes to empty completely and dry out during rest periods. However, in the new PVC perforated pipe with solid couplings, there's always an inch or so of black septic sediment which accumulates in the bottom of the pipe. It's this material that starts to plug the stone media surrounding the pipe. The reason: in order for the effluent to seep out of the pipe, its level must rise up the sides of the pipe to the holes where it can seep out and be absorbed in the soil and purified by more bacteria.

I find the best answer to this problem is to install every third pipe with the holes facing down. This way the effluent can reach the surrounding soils and the black goo has no chance to accumulate in the pipes.

Subsurface Disposal Systems

You have a choice in the matter of your subsurface disposal system, a high-sounding name for the area where the effluent from your home system ends up. The decision will probably depend on your soil conditions, the number of people in your house, the size of your lot, and possibly the building codes in your community.

The *leach field*, sometimes called the *disposal field*, is one of the more common systems. This is a series of trenches approximately two feet deep, two and one-half feet wide, and four feet in between. With two-foot trenches this puts the pipes six feet apart. The width of the trenches is determined by the size of the bucket the excavator happens to have on his backhoe that day. The length of the trenches is determined by the size of the tank and the soil conditions—explained later in the chapter on New Installations.

The length of any single leach line should not exceed 100 feet. Eight 50-foot lines are better than four 100-foot lines, because there's better utilization of the entire field or bed. The first step, after digging the trenches, is to place about six to eight inches of one-inch or one and one-half inch washed stone in the bottom. The perforated pipe is laid on the stone and then covered with more stone, leaving six or so inches at the top of the trench for backfill and topsoil. The ends of the pipes should never be capped.

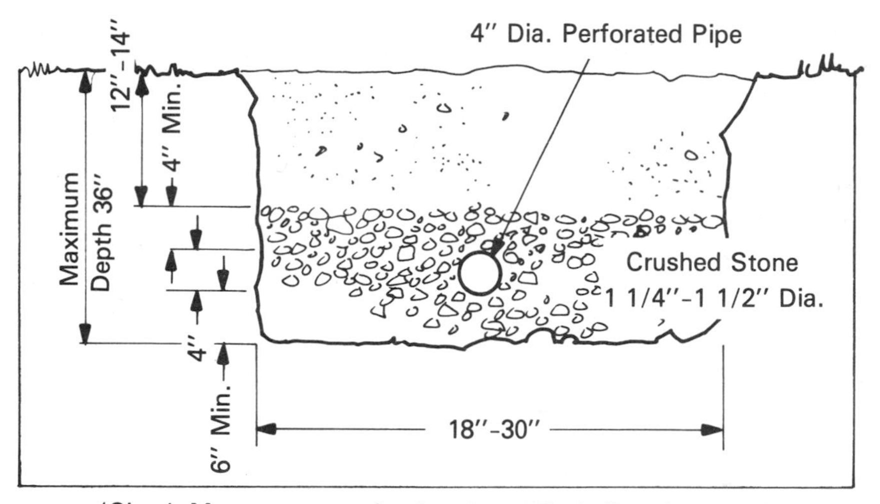

(Check Measurements Against Local Code Requirements)
Pipe Grade 0"-2" Per 100 Ft.

Figure 20. Typical Section Thru An Absorption Trench

Most septic tank installation manuals call for "untreated building paper or hay" to be laid on top of the stone before the backfill is added, so the backfill can't filter down and clog the spaces between the stones. I feel this is totally unnecessary and a waste of money. I don't think it should be done either for the leach bed or the seepage tank, since it has a tendency to seal the soil, thus hampering the capillary action and evapotranspiration. These pipes should be laid as level as possible. (The *maximum* grade for these pipes should be two inches per 100 feet.)

A *leach bed* is exactly the same as the leach field, but instead of being trenched, the entire area is excavated about two feet. Usually this area is about thirty by thirty-five feet, producing a bed of approximately 1,000 square feet.

The leach bed has the ability to accept the same amount of effluent as a leach field of the same square footage. However, I believe the effluent has easier lateral mobility in the leach bed since there is a broad layer of stones. Because, more stone is used, there is also more air contained in the bed, which promotes better secondary treatment. The reason is that the bacteria working on the effluent are aerobic, or oxygen-loving, unlike the anaerobic bacteria on the septic tank, and the bed provides a better environment for them to work in.

This type of installation requires more stone, but it can be built more quickly than a leach field, and is often cheaper to construct.

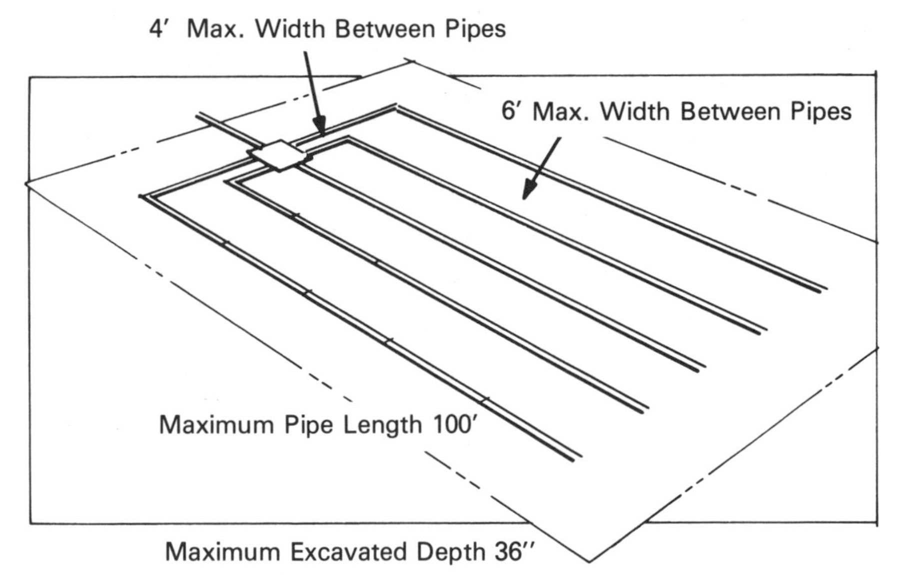

Figure 21. A Typical Leach Bed

Deciding on the size of the leach field or bed can be done by following some general rules, based on the fact that some soils absorb water quickly, and others more slowly. In average soils an application rate of *one gallon* per square foot per day is recommended. While there's no relationship, necessarily, between the size of the septic tank and the amount of effluent flowing from it, the broad rule is to have a square foot of leach field or bed for each gallon held by the tank. Thus, if you have a 1,000-gallon tank, you should have a field or bed covering 1,000 square feet.

Most soils are average. If the soil is less permeable, however, the rate of application should be reduced to three-fourths of a gallon per day, or about 1,500 square feet for a 1,000-gallon tank. In very dense soil, reduce the rate to one-half gallon per spare foot per day, or 2,000 square feet of seepage area for a 1,000-gallon tank.

Curiously, in very permeable soil, such as gravel, the seepage area should be larger than "average" if there's a water supply to protect. The reason is that the effluent will flow through the gravel and into the water system before bacteria have a chance to work on it. By spreading the effluent through a larger area, the flow will be slower, thus giving the bacteria more opportunity to act on the effluent. Sometimes just enlarging the seepage area is not enough, and changing the existing soils to denser ones may be required to slow down the percolation rate.

If you're at all concerned about your water supply, get professional advice!

When choosing between the leach field and the leach bed, I'd recommend the latter whenever possible. I've had good success with it.

Seepage Pit

In good permeable soils a *seepage pit* is often used in place of the leach bed or field. Effluent runs from the septic tank into the pit, then seeps through the pit's earth bottom and porous sides.

There are two kinds of seepage pits, the homemade variety and the "store-bought" one. The homemade type works very well and is less expensive than the precast concrete models.

If you decide to build one yourself, you must first have a hole. It should be about eight feet on each side and eight feet deep and will require approximately ninety concrete building blocks, each measuring 8″ × 8″ × 16″.

The easiest way to build the pit calls for two people and a twelve-foot 2 × 10 plank. One person stands on top and the other in the hole. The person on top slides blocks down the plank and the person in the hole lays the first course of blocks in a circle with the holes facing out. The second

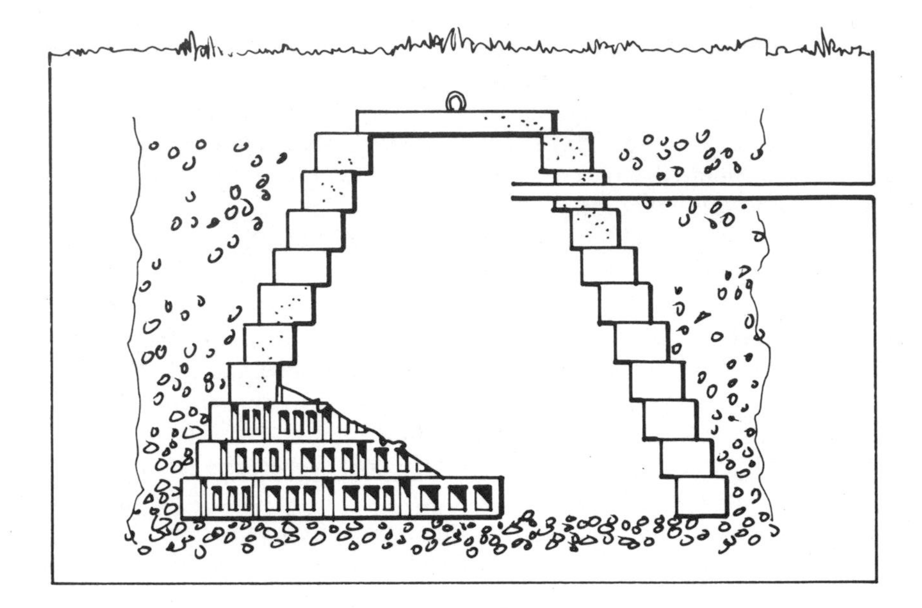

Figure 22. Section Thru Seepage Pit Made From Concrete Block

course is laid with each block pulled in slightly, and with one less block in the circle. Continue in this way with each course, so the diameter of the structure gets smaller, and the finished product looks like an old-fashioned beehive.

Once the first course is laid, the whole thing should only take about twenty minutes to build. Finally, the outlet pipe from the septic tank is installed, and a cover placed on top. Washed stone is placed around the outside of the block wall, to within a foot or so of the top, and then the area can be graded off with topsoil.

The second type of seepage pit is made of precast concrete. This resembles a septic tank in shape, except that the bottom is open, the sides are perforated for seepage, and there's only one cover in the center of the tank. Seepage pits are designed to be used in good, permeable soil, such as sand or gravel. To install one in any dense soil, such as clay, is a waste of time and money. Whenever possible, I prefer the homemade seepage pit over the precast one. Being larger, it allows for a far greater absorption area, and is considerably less expensive.

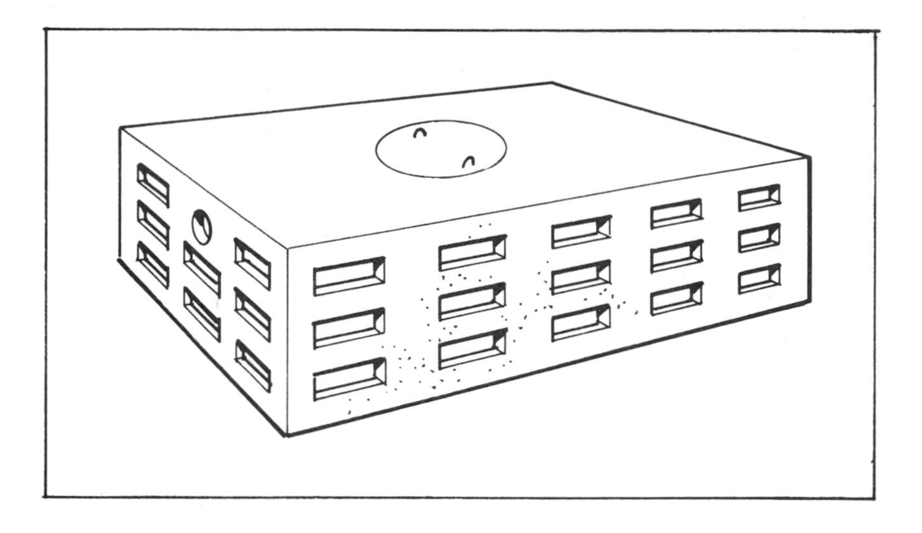

Figure 22A. A Precast Concrete Seepage Pit

There are several ways to use these. If you're in a wooded area, and clearing a large site is not feasible, a series of two or three pits, built so that the overflow from one runs into the next, is practical as long as the soil conditions are good. In this case, every pit except the last should be baffled, so that suspended solids are retained in the pits. In this way, each pit acts as a settling tank, and suspended matter in the effluent is reduced in each tank. Such a system can produce an effluent of 90 to 95 percent clarity, making it possible for the disposal field to last *indefinitely*.

Sometimes a seepage pit is installed between a septic tank and leach field or bed. Not everyone agrees with this method, but it works, and works well.

Here's the reason: at best, the septic tank retains about 65 percent of the total solids received. If it's possible to retain another 20 or 30 percent in a seepage pit, the quality of the effluent will be improved dramatically, which means the leach field or bed will last considerably longer and be trouble-free.

Today the last, and least-used, subsurface disposal system is the *dry well*. It's no more than a hole in the ground, filled with stone, connected to the septic tank by a pipe, and covered with topsoil. Other than the fact that its capacity is less than that of seepage pits, its functions are exactly the same. As the effluent passes over the stones, it is acted on by bacteria that continue the purification process. This type of system is seldom used today, however.

CHAPTER 5

The Distribution Box

When the effluent leaves the septic tank, it travels a short distance in a solid pipe, usually Transite or PVC, to a distribution point, either a box or pipe system.

If a distribution box is used, there are any number of types, shapes, and sizes with any number of outlet holes. Regardless of which type is used, the purpose is always the same, and that is to distribute the effluent evenly to the leach field or bed. Unfortunately, this is virtually impossible. Invariably some pipes will receive more flow than others. Only a "dosing system" eliminates this uneven flow, and for residential systems, it's not necessary.

Dosing or siphon systems are used primarily in larger commercial installations. Basically all they do is collect a specific amount of effluent—

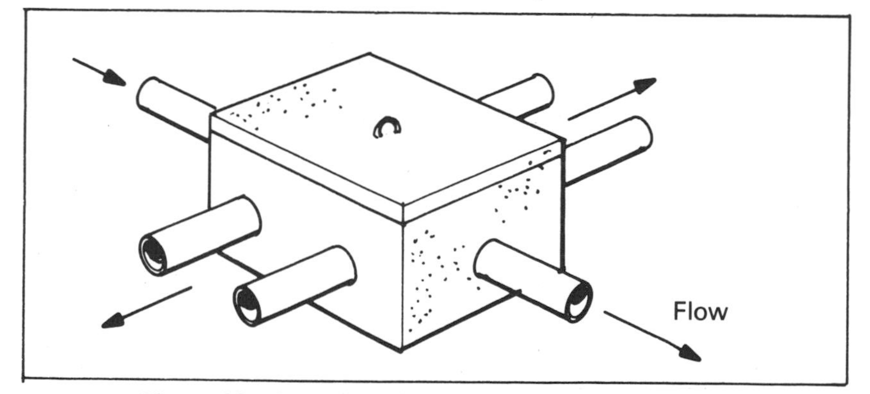

Figure 23. A Residential Concrete Distribution Box

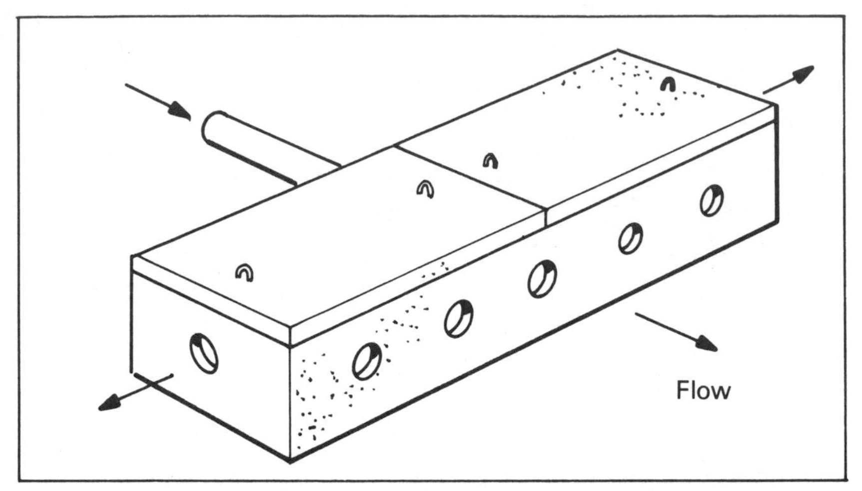

Figure 24. A Large Capacity Concrete Distribution Box

five hundred or a thousand gallons maybe—then quickly release the entire amount to the distribution box. In this way, the field is flooded intermittently and evenly.

An alternate distribution method is to use pipes with four-inch tees and 90 degree bends. Pictured below is a typical leach field using this design.

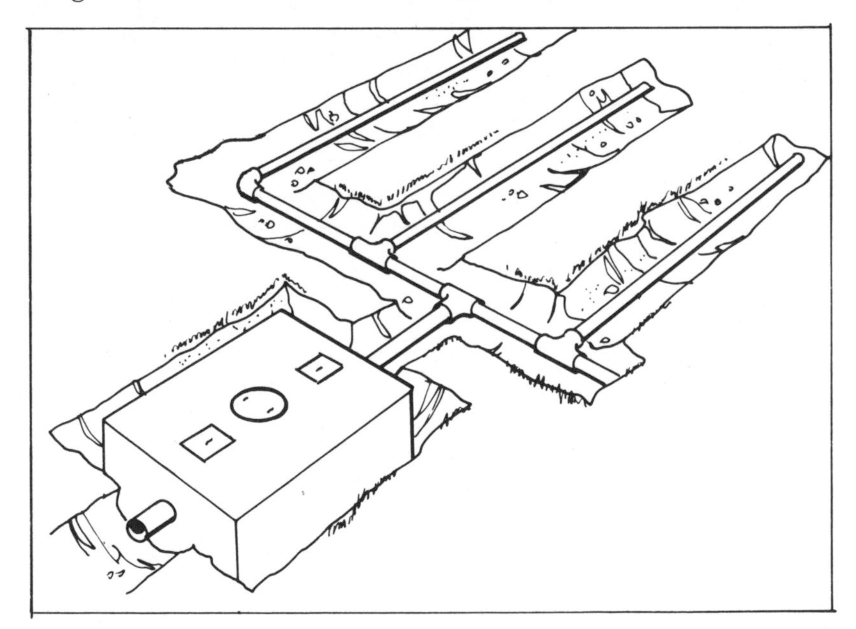

Figure 25. A Leach Field Using 90° Bends And Tees

CHAPTER 6

Other Problems

I can't count the times in the past twenty-five years that we've been at a customer's home frantically digging up his yard in search of his septic tank, when the ever-helpful neighbor has arrived and announced, "Our tank has been in for over fifteen years and *we* haven't had any trouble."

This may be true, but if you live in the country rest assured there are two people you're going to come to know at some time. They are the plumber and the septic tank service man.

As for the neighbor who's had no problems, he's only kidding himself. Because he's never had his septic tank pumped out, the solids are going straight through it and are slowly destroying his leaching system. But for a charge of $50 to $75 to pump out his tank and inspect the baffles and so on, he will ultimately—and *I guarantee it*—replace his leaching system at a cost of $1,000 or $1,500, maybe more.

People who have gone through the expense and aggravation of replacing a leach field and putting the yard back to its original shape don't have to be sold on the idea of periodic, or better yet, scheduled pumpings. A $50 to $75 bill every few years is cheap insurance.

Now, let's look at some of the most common problems associated with country plumbing and septic tanks.

Toilet Backs Up

When the toilet backs up, there is a quick test to determine where the trouble is. Run water in a lavatory, sink, or tub. If these drains work, it indicates that only the toilet is plugged.

The first step is to try the toilet plunger. Fit the rubber head of this plumber's helper over the hole at the bottom of the bowl, then pump with a slow rhythm, lifting it occasionally to reverse the pressure and possibly bring the obstruction back into the bowl. If the water rushes out of the bowl, you've been successful.

But don't pat yourself on the back just yet.

Sometimes you can clear the stoppage with the plunger, only to find, a day or so later, that the same thing happens again.

This indicates there's an obstruction in the toilet trap. It could be a comb, tooth brush, bar of soap, anything. It forms the beginning of a dam to block more debris, which plugs the toilet. Using a plunger clears away the other debris, but it doesn't remove the source of the trouble.

If this happens, or if your original efforts fail to remove the obstruction, it's time to try the toilet auger. This tool consists of a flexible cable, a handle, and a tube that encloses the cable and guides it properly around the big bend in the toilet outlet. This device can either break up the obstruction, hook into it and pull it back out, or force it ahead into the sewage pipe where it will continue what we hope is an uninterrupted trip to the septic tank. (See Figure 3, page 7)

If neither of these approaches works, it's time to call the plumber. He may have to take up the toilet to remove the obstruction.

Sewage Backs Up

Messy problems occur when sewage backs up in your pipes, overflowing at the lowest openings in the system, such as a first-floor shower drain or toilet.

My best advice is to call a plumber, who will probably open the septic tank to see whether there's a blockage in the inlet baffle.

If you open the tank yourself, you can see such a blockage very easily. And you can punch it free with an old broom handle, or other handy device. Stand back, because there will be pressure behind the stoppage and some splashing when it lets go. You should see a four-inch cylinder of waste emerge, followed by a gush of water. Sometimes a little more encouragement with a sewer rod may be needed.

There may be several causes for this trouble. Check for them. Here they are:

1. Look for a sag in the inlet pipe just outside the tank. A sag is indicated if the end of the pipe in the tank tilts up a little. This problem

at the tank end

happens most often when Orangeburg or PVC pipe has been used as a connection between the house and septic tank. It occurs because the back-fill was not compacted around the pipe when the tank was installed. This is simple to fix. Dig up the pipe and replace it.

2. Check to see if the entire pipe has moved out of the tank. This happens when liquid leaks out around the pipe and washes the fill away, leaving a void. The pipe then moves down the outside of the tank. Just dig it up and replace it. In both cases, replace the original pipe with cast iron or Transite pipe.
3. See if the inlet pipe extends too far into the tank. Sometimes we find them pushed all the way up to, and against, the baffle. This is no big problem once you find it. First determine what kind of pipe it is. If it's Orangeburg or PVC, break it off inside the tank. If it's clay tile or cast iron, don't try to break it off, since it may split lengthwise outside the tank, and cause worse problems. The best procedure is to dig it up outside the tank and have a plumber fix it properly.
4. In some cases, sewage backup may be caused by an obstruction in the house sewage line leading to the septic tank.

at the house end

The way to approach this is *cautiously*. You'll be opening the pipe through one of the cleanout plugs. First, put a large bucket under the plug. Then loosen the plug, very slowly. When liquid begins to run out, stop turning, and let the liquid fill the bucket. If the bucket fills, tighten the plug, dump the bucket, and start again. When no more liquid pours out, it's safe to remove the plug.

Sometimes it's impossible to remove the cleanout plug with a pipe wrench. What I do is get a three-pound hammer and a large cold chisel, start at the top outside edge and wallop the plug at a counter-clockwise angle (the same direction you would unscrew it). Then try again at the ten o'clock position, the eight o'clock position, and so forth all the way around the outside edge of the cap. Keep after it. It will eventually loosen. When you replace the cap a little grease will make future removal a lot easier.

Attack the obstruction with a plumber's snake.

Full Septic Tank

If, when you check the septic tank for obstructions, you find the tank is full, you have a different problem. Ordinarily, when the tank is operating well, the level of liquid inside is governed by the outlet opening, which is slightly lower than the inlet. If you use a gallon of water in the house, that gallon flows into the septic tank, and a gallon of effluent flows through the outlet and on to the leach field.

If the tank is full to overflowing, obviously liquid is not getting out of the tank.

Here's the way to check for trouble:

Examine the outlet baffle. Sometimes, but not too often, it can be plugged up at the bottom of the baffle. In this case the outlet pipe will be visible.

After clearing that stoppage, check to see if the outlet pipe extends too far into the tank. Again, correct length is necessary. How do those pipes get shoved in too far? Remember, when the tank was installed, the pipe was just shoved in. Wherever it stopped it stayed. Just a case of poor workmanship.

But suppose you can't see the outlet pipe and the tank is overflowing? Two things are possible.

(a) The leach field has either failed or it's saturated and will not accept any more liquid, or
(b) There's a stoppage in the outlet pipe to the leach area. How do we know which one?

First we can check through the cover of the distribution box. If that box, too, is filled to overflowing, there's an indication the leach field has failed.

Second, if there is no distribution box, we can rod the pipe to see if there's a stoppage or a broken pipe section.

And if this doesn't work, we pump the tank. If the effluent runs back into the tank from the outlet, and it usually does for a considerable time, we've discovered that the problem is in the leach field.

Does this mean a new leach field? Not quite yet. Sometimes, and only sometimes, by pumping a tank and putting less water into it for a time, you can get the leach area to dry out and again accept liquid. But don't depend on it. The only safe solution here is to get the best advice possible from someone who knows local conditions. Then follow it.

If the leach system has failed and the tank is a vintage model, replace both the tank and the field. A new field with an old tank is no bargain.

Many times this problem is caused by an unusually wet spell, particularly in the spring when melting snow and spring showers saturate the soil.

Whenever I tell people their leach field has become plugged with solids, they ask, "How did those solids get through my tank?"

The answer is: two ways. First, and most common, there's pure neglect. If you look at Figure 17, (page 22), you'll see an accumulation of solids between the two baffles. There is virtually no head space; no area above the water line allowing the tank to accept large amounts of water from a bathtub or washing machine rapidly unloading twenty to thirty gallons of water. Similarly, the collection area between the baffles is greatly reduced. This means that sewage from the house is funneled straight through the tank, and the solids don't remain long enough to be digested.

The second problem is faulty construction. Water can act like a solid, and if the pitch of the pipe running from the house to the tank is too steep, water enters the tank at an excessive velocity. At about eight pounds per gallon, a toilet flush produces a forty-pound slug of water. Drop that down a four-inch pipe, and the impact at the tank is more than enough to keep the contents stirred up. With solids floating at all levels in the tank, some will flow past the outlet baffle and into the leaching system.

This is why the grade of the pipe and the selection of the tank site are critical. The slope of pipe to tank should be about one inch in ten feet, and whenever possible, the tank should be about ten feet from the house.

If there are facilities in a basement, many installers will put a tank more than ten feet from the house. Too great a distance to the tank is almost as bad as too much drop. The solids will either be stranded in the pipe, or they will accumulate in the inlet baffle, causing a blockage. Either one can cause a recurring problem. But at least they won't spoil the leaching system.

Slow Toilet Flush

Here's a common problem in country plumbing. The toilet flushes slowly, or a sink doesn't drain as quickly as usual during the day. Occasionally there's a gurgling sound coming from the depths of the sink.

But you're hopeful the trouble will go away, because next morning everything works just fine. Several days go by, and the trouble reappears, the grumbles grow louder in the sink, the runoff of water is slower and slower.

You think, "It'll take care of itself."

It doesn't. After several days have gone by and numerous boxes of Septic Wonder have been force-fed to the tank, you are back to Square One.

"Call the plumber, Alice."

What's happening down there?

Simple. Overnight the liquid manages to seep out slowly. This leaves space in the sewer pipe, enough to accept limited use in the morning with no sign of trouble.

If there's a good side to this problem, it's that it gives you fair warning. Those gurgles and those slow drains are telling you something: a backup is imminent.

When it arrives, you'll know what to do. I hope.

When all of the facilities back up like this, it's sometimes difficult to identify the real problem.

You uncover the tank and find that the liquid level is where it's supposed to be. The solid accumulation in the center section is only an inch or two thick, but at the inlet baffle there's a large blockage. So, with a poke or two of the old broom handle, you open it up. After the pipe has emptied into the tank, you flush a toilet to be sure the line is open.

Everything looks fine. You replace the cover, and with a great feeling of accomplishment you cover up the tank.

Hold on. The real problem still exists. It's the leach field. But, you say, the outlet pipe was okay when the tank was opened.

This is what happens.

First, the leach field became saturated and would not process any more effluent. If liquid can't leave the tank, obviously more can't get in. So the liquid builds up in the tank, then starts to back up in the inlet pipe. Material coming from the house gradually starts to build up at the inlet baffle, then slowly backs up toward the house.

But the toilets worked in the morning, you say. This is because some liquid seeped past the stoppage overnight when a little effluent was handled by the leach field. This made a small space in the pipe which was filled in the morning.

The next step is for the pipe to become totally plugged, and the backup seems here to stay. In the meantime, however, water continues to seep slowly out of the tank, and by the time the tank is uncovered, all indications are that it's working normally except for the blockage at the inlet baffle.

This situation has fooled me several times. How do you know what to look for? The first clue is that the facilities worked on a limited basis this morning. Second, after the tank is opened, you can see the liquid level has been up to the top of the tank, a sure sign that the leaching system was not able to handle the load. Ordinarily the liquid level leaves a ring around the inside of the tank, a ring that's very visible. Above that ring the tank is dry and shows signs of rust. This is always good evidence of the condition of the leach field.

This could be a seasonal thing, caused by unusually wet weather. If not, you had better plan to rebuild or extend the leach field.

This section has covered *all* of the possibilities associated with a backup, and they are:

1. Simple stoppage in the toilet bowl.
   Use a plunger or toilet auger.
   Call a plumber.
2. Blockage in a pipe between house and tank.
   Try rodding from house.
3. Blockage in inlet baffle.
   Uncover tank and punch out.
4. Tank needs pumping.
   Call a septic tank service company.
5. Outlet baffle plugged.
   Punch clear.
6. Outlet pipe plugged or broken.
   Check with sewer rod.
7. Leach field saturated or plugged.
   Call a reputable excavator.
   The field may have to be extended or replaced.
8. Any combination of the above.

When the Sewer Rod Doesn't Work

If you have a stoppage between the house and the tank, and you've tried a sewer rod with no success, here's a little tip:

What's probably happening, and it's happened to me many times, is that the stoppage is so soft that the sewer rod—the "snake" as it's called—just goes back and forth through the stoppage, doing nothing. First remove the cleanout plug in the basement, then push the sewer rod from the tank to the house.

When the rod reaches the cleanout hole, pull it out a little and tie a rag on the end of the snake. Make it good sized, but not so big it will get stuck in the pipe. Then go out to the tank and pull the snake out. You may have to repeat this operation several times. *Always* try to pull to the tank.

Finally, if a cleanout is not available, just tie a rag on the snake and push in and out from the tank. Don't get discouraged. It will come eventually.

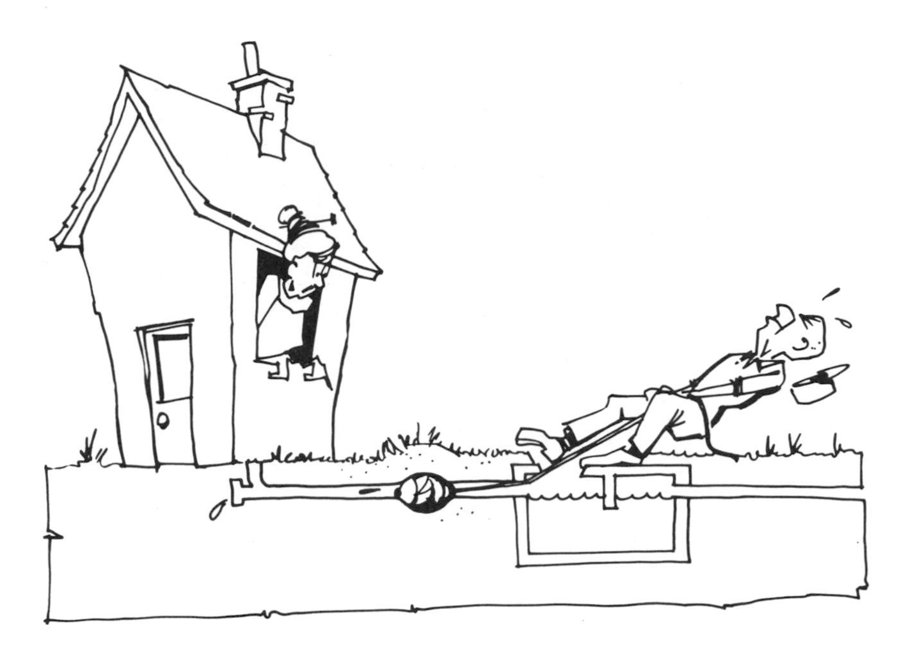

between 'em both

After you've cleared a stoppage, always flush a toilet to be sure the line is open before covering the tank. It could save you digging it up again.

By now you must be wondering, "What else could happen?"

Well, if you have a steel tank, the cover could collapse or the baffles could rust out and fall off. Since replacement covers are not available, and having a steel replacement cover made is extremely expensive, what do you do? Here's how to fix them.

You replace the steel cover with wooden planks. Use either spruce or hemlock. Use 2 × 8s or 2 × 10s, or even better, bridge plank. What's bridge plank? It's the kind of timber used on bridges, usually 3 × 10 by any length. Bridge plank is available at most rural lumber yards.

Now, there's a way to do this that can save you considerable digging. The accompanying illustration shows you how to place the planks:

You'll notice the center plank is placed over both the inlet and outlet baffles so that, in the future, should you have to pump the tank or clear a baffle, only the center plank has to be removed. Do not put tar paper or

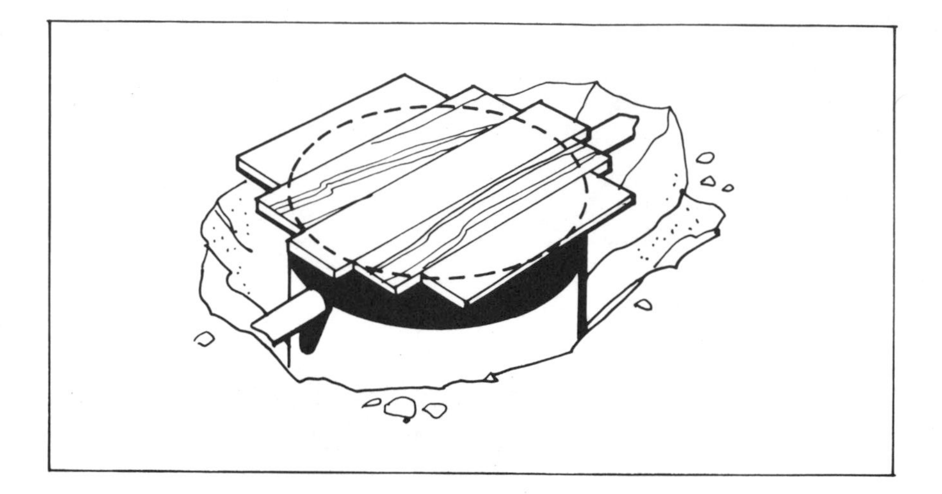

Figure 26. Typical Placement Of Planks To Replace The Steel Cover Of A Septic Tank

other waterproofing material on top of the planks. It will only hold moisture and shorten the life of the boards. Just backfull.

In metal tanks, rusting occurs only in the area above the water level. This includes the cover and the tops of both baffles. Replacement baffles are also unobtainable, so something will have to be fabricated. Look around for an old five-gallon pail, plastic if possible. Cut out the bottom of the bucket, then cut the remainder in half. Now you have something that looks like the two curved pieces in the illustration. (Figure 27)

You'll need some way to hang them up inside the tank. I usually bend two eight-penny nails as shown. Then I punch holes in the upper two corners, put the nails through the holes, and hang the new baffle over the old one and over the rim of the tank. Don't worry about its moving; the cover will hold it in place.

That's fine for the inlet baffle, but the outlet baffle should be six inches longer. Sometimes only the top part of the original baffle will be rusted away, and since the outlet side is the important one, try to find another pail, so the two sides can be attached as shown in Figure 28.

The job is now finished. As you replace the cover, you may notice small holes rusted through it, particularly over the inlet baffle. This is not un-

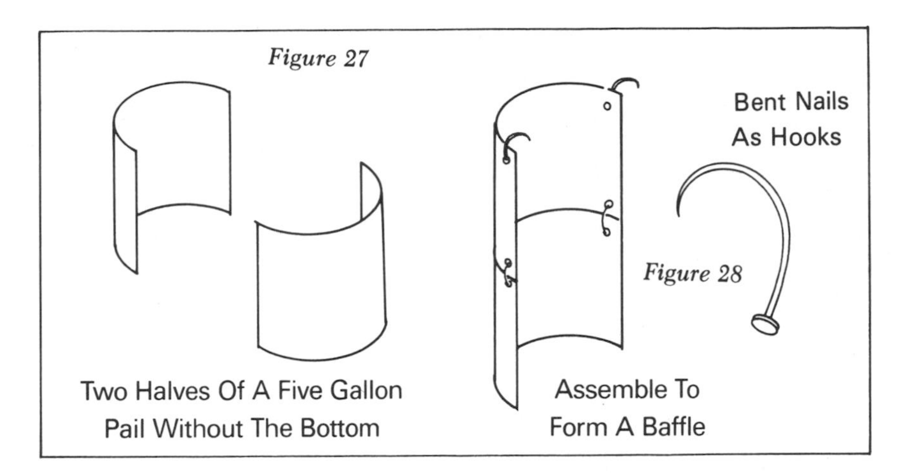

Figures 27 And 28. Baffles Made From A Pail

usual. Just put a piece of tin or tarpaper over the holes in the cover and backfull. If there are any clips on the tank, you don't have to bend them back down. They're there only to keep the cover on during shipping and installation.

Tree Roots

Tree roots get into clay tile more often than other types of pipe. This becomes a common problem when clay tile is used between the house and the septic tank. The reason is that a perfectly tight joint is not always made when the pipe is cemented together. Also ground movement can cause the tile or cement to crack. The result is a tiny hole—big enough for a small tree root to enter. Once inside, this little root takes on added vigor, and with all that added nourishment it resembles Topsy in *Uncle Tom's Cabin*—it just grows and grows until it plugs the pipe.

Sometimes, if the pipe is not completely plugged, it's possible to kill the roots with an application of copper sulfate, sometimes called Blue Vitriol. This is available in most hardware or drug stores, and a five-pound box is all you need. The best method of application is to dissolve half the box in a bucket of hot water and flush it down the toilet. Repeat this with the remaining half the next day. If this doesn't work, don't be disappointed. The root stoppage may be too large or the solution may not be reaching the roots.

The next step calls for rodding with a municipal sewer rod or an electric snake with a root cutter on its end. This tool is especially designed to cut and remove roots. Check the Yellow Pages under, "Plumbing—Sewer and Drain Cleaning" for this service.

For a job that won't have to be repeated, the only solution is to dig up the pipe and repair it. All other methods leave the opening, and some day another tiny root will find it. Also, continued introduction of chemicals into the septic tank is not recommended. It will halt or slow the bacterial action in the tank.

Winter Problems

Let's say it's the middle of the winter, and you have a frozen sewer pipe. Some people tell me they've flushed rock salt down the toilet and it worked. This may be so, but usually when people get through fooling around and call me, they have a situation that only a steam generator can help. A generator with a garden house attached to a flat sewer rod pushed through the cleanout in the basement always works. Since many plumbers and garages have these machines, and they're portable, they're quite handy as well as available.

The bad part about using a steam generator to open a pipe is that after the pipe is cleared, you have to leave a small amount of water running through it or it will freeze again. The same is true in any water pipe.

Freezing usually occurs when there is no snow covering the pipe or tank, or where vehicle traffic over the pipe drives frost into the ground. If you have shallow sewer or water pipes, don't drive or even walk over them, because even foot traffic can drive frost downward.

If you have to dig up a frozen pipe, an air compressor and jack hammer are usually needed. Again, if it's not protected afterwards, the pipe will freeze again. To prevent refreezing, use 2 × 10 planks to build a three-sided box around the pipe, fill the box with insulation, and backfill the whole thing with sand. (Figure 29)

Another procedure for protecting a water line that's close to the surface is to enclose it in a four-inch solid sewer pipe, and bury it as deep as possible. The air surrounding the pipe will insulate it from freezing.

A septic tank will rarely freeze in the winter. There's a flow of warm water into it most of the time, and the digestion process generates heat, so much in fact, that you can often see where your tank is during the winter because the snow on top of it melts.

Septic tanks can fail at any time. Should you have a problem and have to dig up your tank in the winter, you'll find that after you shovel the snow off, digging is relatively easy.

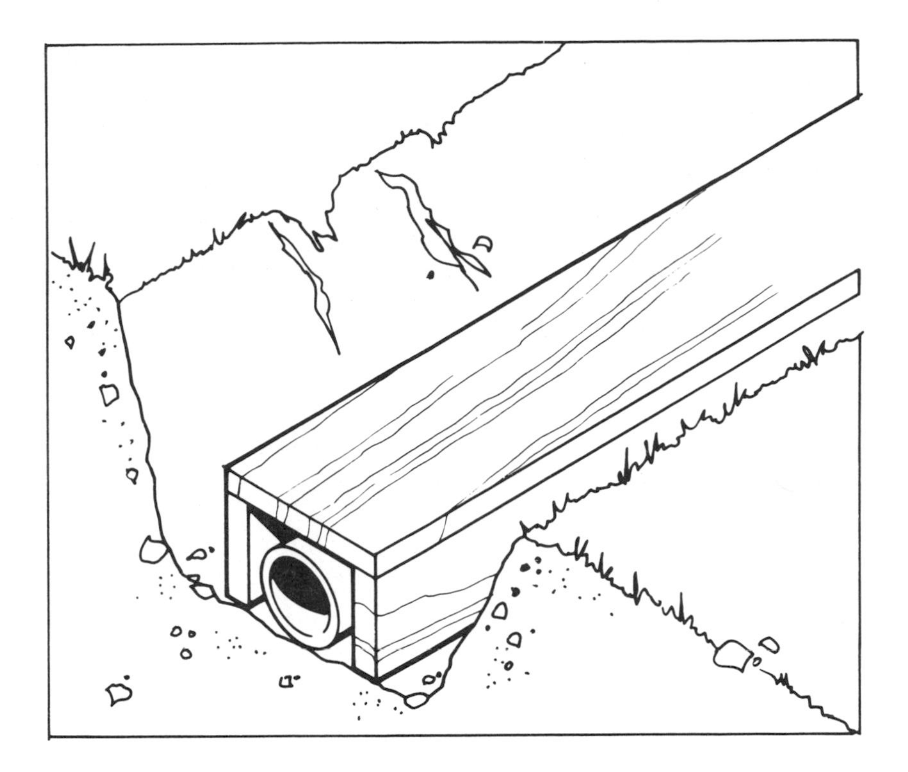

Figure 29. Boxing A Pipe To Prevent Freezing

However, keep in mind that the dirt has to be replaced the same day or it will freeze in chunks, and putting it back over the tank will be impossible. If for some reason you *can't* backfill that day, spread a bale or two of hay over the top of the tank, and it will not freeze.

If you've uncovered the tank and find it needs pumping, but for some reason the pumper can't come, or can't get close enough to pump it, (which happens), a temporary solution is to shovel out some of the solids floating at the top of the tank. I've never seen a tank we couldn't make work for a month or so until we could return and pump it properly.

Material that is shoveled out can be buried in a shallow trench or covered with chlorinated lime or snow. At freezing temperatures there will be no odors, and by spring it will have disappeared anyway.

If you have a dog, tie him up. There's nothing a dog would rather do than roll around in something that smells bad, then insist on coming in the house.

If a leach field fails in winter, the only solution is to install a temporary seepage pit, until the system can be fixed correctly.

The Field That Fails

What happens when a leach line fails? As you can see from Figure 30, only the area surrounding the pipe is plugged, while the remaining stone is clean and still usable. This black septic residue will extend anywhere from one to two inches around the pipe, and the thickest accumulation generally is on the bottom.

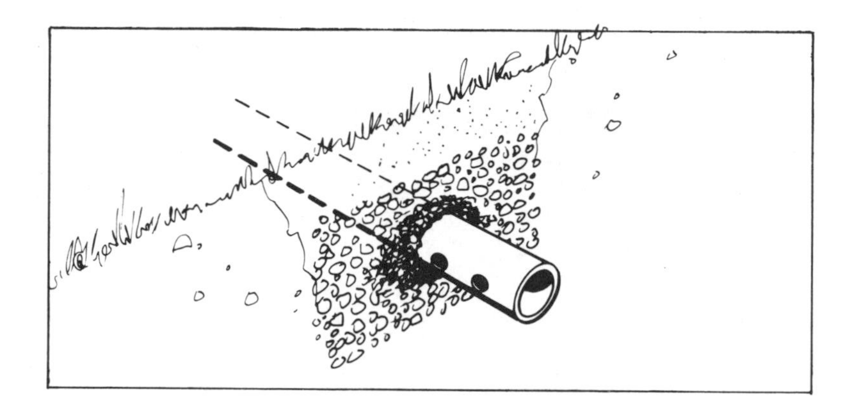

Figure 30. Section Thru A Leach Trench Showing Black Material Surrounding Pipe Causing The Media To Become Plugged. (Notice Clean Stone Remaining)

There is, to my knowledge, no way to clean the residue from stone, either around the pipes or around a seepage pit. I have tried flushing, rodding, and using chemicals in attempts to open up these areas. They didn't work. The only way to correct the problem is to replace the lines and the stone. It is necessary to replace only the stone, not the blocks, around the walls of a seepage pit.

Many environmental engineers require a "reserve area," so that if one leach field fails, there is room to put in another. In all the fields we've rebuilt, we've never abandoned an area. They have always been rebuilt in the same place. We can even salvage much of the stone around the pipes, which is a great savings.

Figure 31. In Memory Of The "Reserve Area"

Properly designed, installed, and maintained, a leaching system should last indefinitely. You should remember that the septic tank is designed to protect the leach system by producing an effluent that will not plug the field. This will happen only if the tank is maintained properly, and pumped out regularly, so that the quality of the effluent remains high.

Leach Field Failures

When a leach field fails, you should determine *why* it failed, then correct the problem so it doesn't occur again.

Disposal systems suffer from the same problems as the septic tank: poor design, installation, and maintenance. In almost every case of a failed leaching system, it's a direct result of one or several of these three reasons. Here are more specific reasons for failure:

A. Poor design:
   1. Insufficient leaching area.
   2. Poor choice of a system, such as a seepage pit in heavy soil.
   3. Septic tank too small.

B. Poor installation:

1. Pipes too deep in the ground.
2. Too steep a pitch in the pipe leading into the septic tank. This permits fast-moving sewage to stir up the tank, and carry suspended solids into the leach field.
3. Too few or no stones around the leaching pipes. They should be completely surrounded by 1¼ to 1½-inch washed stones.
4. Bypassing the septic tank with grey water from the washing machine, tub, sink or shower, allowing it to run directly into the leach field.

C. Poor maintenance:

1. Septic tank not pumped frequently enough, allowing the leach field to receive excess solids which can only plug the system.

It's impossible to state the exact amount of time that should elapse between pumpouts. If the septic tank is small, or if there's a large family using a medium-size tank, the interval can be as long as five years or even longer. A commercial pumper can tell you, when he's pumped the tank, whether you waited too long, or might be able to wait longer the next time. It's safer to be on the side of too frequent pumpings.

When you've found the cause of your problem, the next step is to correct it. This is not always easy as the size and type of system (especially the leach field) is not always known. About all you can be sure of is that it's not working. Sometimes giving the field a rest will cure its ills. This means pumping the tank, then using as little water as possible so little goes into the system. This treatment allows the field to dry out, and sometimes it will again accept effluent. If this works, it's usually because the field has been receiving an excess supply of water during lengthy wet periods, when the snow is melting, or when a toilet ball valve is leaking.

If it doesn't work, it means you'll have to extend or replace the field. And if you do, it's probably best to also replace the tank since it's probably the tank that caused the problem to begin with. If you have a 300-gallon or 500-gallon tank that's old, it should be replaced. A new field with a poor tank, as I said before, is no bargain.

Finally, if the top of the old tank is buried more than two feet below the surface, it's better to put the new tank in at a shallower depth, even if it means some plumbing changes inside the house and an additional hole punched through the cellar wall. The old hole can always be patched. And this will make for a far better system because it puts the tank at the proper depth, meaning the leach field also can be at the proper depth.

Remember, if the top of the septic tank is three or four feet down, excavation for the leach field or bed must be at least four or five feet deep, and that's far too deep for the field or bed to work most effectively.

Odors

This brings us to the problem of odors, which can be troublesome either inside or outside the house.

We'll start indoors. While a septic tank is no rose garden, the odor from sewer pipes inside a house is just awful. It's a smell resembling natural gas, only stronger, and it can drive you out of the house.

Where does it come from? The septic tank and sewer pipes. The gas is one byproduct of the bacterial action in the tank. If the system works as it should, the gas comes back into the house by way of the sewer pipe, then goes up the vent stack (that pipe sticking up out of your roof), and out of the house. Then how does that gas get out of the venting system and into the house?

Basically, two ways, either through a crack in a toilet seal or through some trap that has lost its water seal. Sometimes a floor drain is connected to a sewer pipe. With or without a trap, this is a violation of plumbing codes, and if this type of drain connection is the source of the trouble, it should be changed.

But first, when you discover an odor, try to find out where it's strongest. If it's in the bathroom, it's most likely a toilet seal, particularly if you have an old house.

Where is the toilet seal, what is it, and how does it leak?

First, it's between the bottom of the toilet and the toilet flange set in the floor. Second, it's made out of beeswax, and third, after a while that wax will dry out, permitting gas to escape. Old houses with old floors move with the changing seasons, causing the seal to leak. Just general traffic will loosen a bowl after a while.

This sanitary ring, as it's sometimes called, resembles a large doughnut. It is placed in position under the toilet before the toilet is tightened down with the four little bolts at the base. This compresses the ring and provides the seal. Sometimes two seals are required. Since they are below the water trap in the toilet bowl, the gas in the system can flow up to them and escape through even a small pinhole making life unbearable in the house. A call to the plumber will solve this problem.

It's easier to correct the other odors inside the house. If the house is unoccupied for a period of time, or if a particular shower or lavatory is not used, the water in the traps will evaporate. This allows the gas—and odor—to permeate the house. All you need to do is run a little water to fill up the trap. And that's that.

donut?

But what about floor drains?

There are two types. The first is just a regular trap with a four-, three-, or two-inch pipe going straight down through the floor. It has some sort of a perforate cover. The second resembles the first, except there's a small moat around the pipe. If you turn the cover over you will see a circular section attached to it. This is called a bell trap.

To stop odors in a regular floor drain, just put some water in the pipe. In the bell trap, fill the little moat at the top, and as you replace the cover you will see how it enters the water to form the trap. There's a way to slow evaporation of the water from the drains, and that's by putting a small amount of fuel oil on top of the water every now and then.

All of this should take care of odors in the house—unless of course, something furry dies inside one of the partitions.

Next we move outside. Take a look at the roof and locate a three-inch pipe sticking out of it somewhere. There may be more than one. It's called a vent pipe or stack, and it's the one that lets the gas flow out of the septic tank.

It also does something else. It allows wastes from the house to flow freely through the pipes to the septic tank by venting the fixtures. Without this pipe the house fixtures would drain poorly, and the traps might be drained, permitting gas to flow into the building.

The vent pipe, when properly installed, should be within three feet of any plumbing fixture that requires venting. If you have a long ranch-type house with a bathroom at one end and the kitchen at the other, you may need two vent pipes poking through the roof, unless you have a loop vent system.

A venting problem exists in my own home where the bathrooms are at one end of the house and the kitchen is at the other. Since we have only one vent pipe, every time the washing machine in the basement empties, we have a gurgling sound in the kitchen. What's happening is that waste water leaving the washer is siphoning the water out of the kitchen trap. If the kitchen were vented, this couldn't happen. Some day I'll fix it.

But the vent can become a problem, depending on where your house is located. Remember, it's connected directly to the sewer pipe going to your septic tank, and sewer gas generated in the tank rises out of the vent pipe. You may never have noticed, but the first thing roofers do when they get on a roof is stuff a rag down this pipe. This improves the working conditions considerably.

Have you ever noticed that some days the odor outside is worse than others? Well, the weather has a great deal to do with it. You have just finished a wash, and the weather outside is hot and humid. No wind. As the washing machine empties, the waste water displaces the air on the top of the septic tank, forcing the air and gas back up the sewer pipe to the vent, and finally to the roof.

Being heavier than air, and with no wind to disperse it, the odor settled down around the outside of the house, usually by an open window.

This problem often occurs the day you decide to have an outside barbecue in the back yard. Everything is going just great. It's a nice, sunny day, and the charcoal has finally decided to burn. The beer and gin 'n tonics are serving their purpose, and the traffic to the head is directly proportionate to the liquid consumed, if not greater. And then it arrives—that godawful smell.

Well, folks, the next move is to the front yard. With any luck it'll rain.

As I mentioned earlier, the location of the house has a great deal to do with the severity of the problem. Is it situated in a low area with a hill or trees surrounding it that restrict the wind flow? Is the vent pipe high enough? Since moving the house is not feasible, there are two ways to eliminate this problem.

The first is to extend the existing pipe above the ridge of the roof or any part of the roof that could block air circulation. This will improve the conditions greatly. However, the aesthetics of the house deteriorate with the length of the pipe.

There is a second solution: the installation of a running trap. This is placed in the main sewer line, in the basement just before it leaves the house. The typical running trap shown here is made of cast iron.

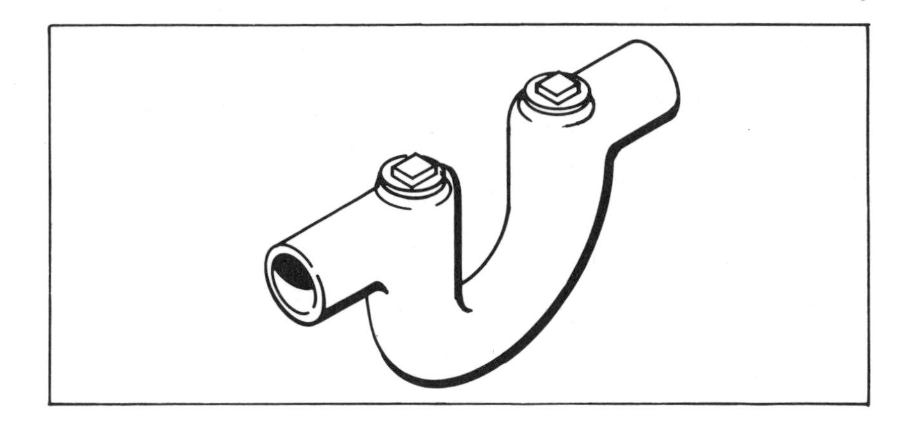

Figure 32. A Running Trap With Two Cleanout Plugs

The trap has two clean-out plugs on its top. These traps always work and are nearly always found in homes on municipal sewer systems to keep sewer gases from escaping via the vent pipe.

But there's a catch. The trap has a tendency to slow the flow of waste water to the septic tank. It also allows grease to build up in the trap area. This only requires removal of the clean-out plugs from time to time to punch the stoppage free. Since people are generally careless about grease anyway, there is no avoiding this problem. Since nothing is perfect, at least the odor from the vent pipe will have been eliminated. This may be the lesser of two evils.

Now comes the universal problem, a wet area in the yard. Sometimes it's black or grey, and always it's accompanied by odor.

This situation usually occurs in the first few years a new house is occupied, and the first thing people think about is having the tank pumped out. Pumping in this situation is only a waste of money. It won't correct the problem. The cause of sewage surfacing is a poor leach field. Initially the leach field may have been too small. But even if the field is adequate, it may have been improperly installed, which is most often the case.

Nine times out of ten the pipes have been laid with too much pitch or grade. Result: all the effluent ends up at the end of one or two pipes. No matter how large the field is, 80 or 90 percent of the area is not being used. The remaining 10 to 20 percent can't possibly handle the load. As the effluent collects in one area, the ground becomes saturated and "sewer sick" because the bacteria can't handle it. Consequently it develops an odor — a very, very strong odor.

Sometimes the problem will be slightly different. When you uncover the distribution box it's evident that one or two lines are getting all of the effluent. If this is the case, it may be possible to plug the pipes with an old rag, thus directing the effluent to the unused pipes. After letting the plugged pipes and wet area dry out for a while , you can redirect the flow in the box by using a small amount of hydraulic cement, available at most hardware or building supply houses. The easiest way to judge where to put the cement is by running a garden house into the distribution box, so you can see the flow pattern.

In either case, if the problem persists, it will be necessary either to extend the field, if you have the land, or to rebuild the original field properly.

Meanwhile, if the odor persists, there's something you can do for temporary relief, at least until the situation can be corrected. Mix a cup of Clorox in a bucket of water and sprinkle this over the area. Swimming pool sanitizers, liquid or dry, will also work.

Many times contractors use seepage pits because they're the quickest and cheapest system to install. Seepage pits should be used only where soils are suitable, such as gravel, or wherever the absorption rate is good. If the soil is more dense, the absorption rate will be less, and a leach field or bed must be used.

In all fairness, the contractor may not be entirely at fault when a leaching system fails. He may have subcontracted the sewage disposal in-

the ripe time

stallation and so is unaware of what was installed. But if he's reputable he should know. At least part of the blame is his.

By now you must agree, the privy was not all that bad.

CHAPTER 7

The New Installation

A look at the many problems that occur in a septic system should convince you of one thing: it should be carefully planned and constructed so that most of those headaches never arise.

If you plan to build a home, there are several steps to take before you start construction of the septic system.

First, contact the local planning commission, if your town has zoning, or the local health officer, and ask about the sewer ordinances and regulations. You'll have to follow them, so it's best to know about them as early as possible.

Second, arrange for a backhoe to dig at least one test hole. This hole should be at least eight feet deep and it should be located in the area of your planned house and disposal area. It will tell you all sorts of things: the type of soil, which determines the type and size of the leaching system, and whether or not ledge or subsurface ground water is present.

All the information you obtain from your test hole is very important. After you have started construction, you don't want any surprises. As long as you know what's in the ground, you or your architect can plan accordingly.

The Perc Test

Some towns and cities require that "perc" (percolation) tests be taken before a new septic system is installed. This means that a professional engineer or some other approved official will dig a small hole about 24 inches deep, pour a bucket of water into it, then measure the rate of absorption with a ruler and watch.

The results will indicate the soil's ability to absorb liquids, and this in turn will determine the size of the leach area. Naturally, the slower the absorption rate, the larger the size of the leach field. All unsaturated soils absorb water, some more rapidly than others. By using modern septic installation techniques during site preparation the land can be prepared for almost anything. All it takes is money.

Remember that hole we had dug with a backhoe? Still more information can be gained from it. Let's suppose we run into water at five feet. How is this going to affect the cellar hole? As long as you know the water table is five feet down, your architect or contractor can plan accordingly.

The system most commonly used in this instance is a "curtain drain." It resembles a leach line, but instead of allowing liquid to seep out, this pipe collects ground water and carries it away. The depth of a curtain drain can be figured according to the depth where water was struck when the eight-foot hole was dug. (Figure 33)

This installation is used most commonly when a house is built on the side of a hill. The curtain drain is installed around the uphill side of the house. Footing drains may also be tied into the system. All of this helps isolate the leaching area, and lowers the water table, thus reducing any chance the leach field will fail because of the high water table.

This preliminary period is also the time to find out what kind of water your property can produce.

Recently a friend of mine drilled two 500-foot wells in his front yard, and got nothing. Later, when he decided to put a garage behind his house, the backhoe dug out two buckets and struck water at three feet.

Such problems aren't unexpected in country plumbing. But you can often find out a lot before you start drilling.

Call a local well driller and ask about either shallow or deep wells. If the drilling firm has experience in your area, the drillers can give educated guesses as to what results you will get. Contact your plumber for advice, and neighbors to find out what types of water systems they have.

Keep in mind that deep well water is frequently very hard and may require iron removal or softening. In some cases, the water may have to be aerated to remove odors. Spring water or shallow well water, on the other hand, is generally soft and requires no treatment.

Developing a spring is quite inexpensive compared to a drilled well of any depth, so if this type of water is available, I recommend you use it. Should it fail, you can always drill a deep well.

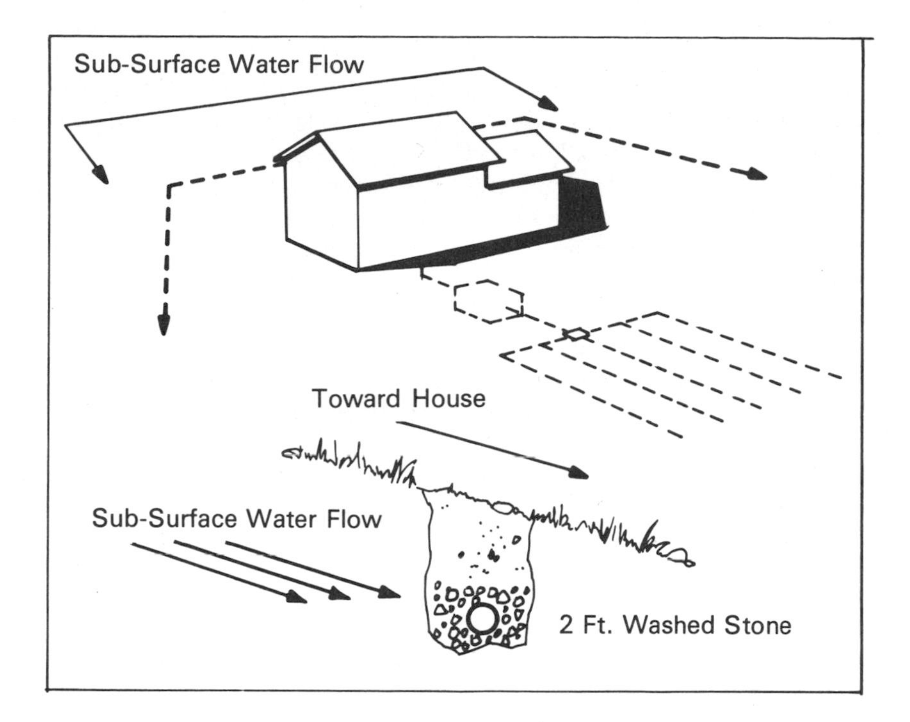

Figure 33. A Typical Curtain Drain With Cross-Section Thru Area

For best results, a septic system should be planned from the house to the disposal area and not the other way around. This way the field, tank, and distribution box can be placed at the proper depth for most efficient operation.

Let's look at some numbers. Remember, the following measurements may vary from the code in one town to the code in another. Use the distances dictated by the building code in your area. No matter which type of disposal area type you select, the principles for installing the system will be the same.

Today, most one and two-bedroom units are either in a condominium or in multiple housing complexes. These usually have community or municipal sewage disposal systems. Since we're concerned with individual rural systems serving three and four bedroom units (that constitutes the largest percentage of homes being built today), these are the figures we use.

For each bedroom allow 250 gallons of water per day. Figure that a three bedroom home should have at least a 750-gallon septic tank, a four bedroom house a 1000-gallon septic tank, and so on.

Now, the square footage of a leach field (bed) should never be *less* than the septic tank capacity (i.e. 1000-gallon septic tank = 1000 square foot leach area). This general rule applies to a large range of soil conditions.

However, should the soil be less permeable, a 50 percent increase in the leach area may be necessary. So a 1000-gallon septic tank might need a 1500 square foot leaching area. In "tight" soils we've sometimes used a 1000-gallon septic tank in conjunction with a 2000 square foot leaching area — a 100 percent increase over normal leach-field size.

What does all this tell you?

The number of bedrooms determines the septic tank size, of course. The permeability of the soil controls the size of leaching area. Thus it's possible to have a small septic tank and large absorption area. But, any tank with less than one square foot of absorption area per gallon of septic tank capacity is not recommended.

No matter what the size of the leaching area, it can be dug in a variety of formations. For instance, a 1000 square foot area could be five eighty-foot trenches that are two and one-half feet wide. (Again, the width of the bucket on the backhoe determines the width of the trench.) If you select a leach bed, the excavated area could be thirty by thirty-five feet, twenty by fifty, or any combination that measures approximately 1000 square feet.

Recommended depths for the trench vary from code to code. One suggestion is that it be no more than twenty-four inches deep. Any deeper, and the action of aerobic bacteria may be slowed. Trenches can be eighteen to thirty inches wide, and should be at least four feet apart. The individual pipes should be no more than 100 feet long.

When non-absorbent soils like heavy clay are dug from the trench or bed, they should be trucked away, and the trench backfilled with sandier soil.

The only way to dispose of the liquid wastes from your home is to run *everything* through the septic tank into the leach area. That's my strong recommendation, anyway. If you bypass the septic tank with so-called grey water — from the lavatory, shower, or clothes washer — and run them straight into the leach field, the grey water will eventually spoil a seepage system.

They contain a high percentage of soap fats, hair, lint, body oils, shampoos, and bath oils, that will combine to form a scum which coats the sides of a dry well and plugs the pores of the soil in any disposal system. Anyone who's had an opportunity to clean out a bathtub drain will readily understand this. Since the septic tank traps most of this material, and keeps it out of the leach system, *all waste waters should go to the septic tank.*

One last thought: the most successful and inexpensive way to install a subsurface disposal system on a side hill is to bulldoze a flat area where the field can be installed properly. Even if you have to prepare *two* levels, or terraces, this is still the best choice, since it's both difficult and costly to build a leach field on a sloping site.

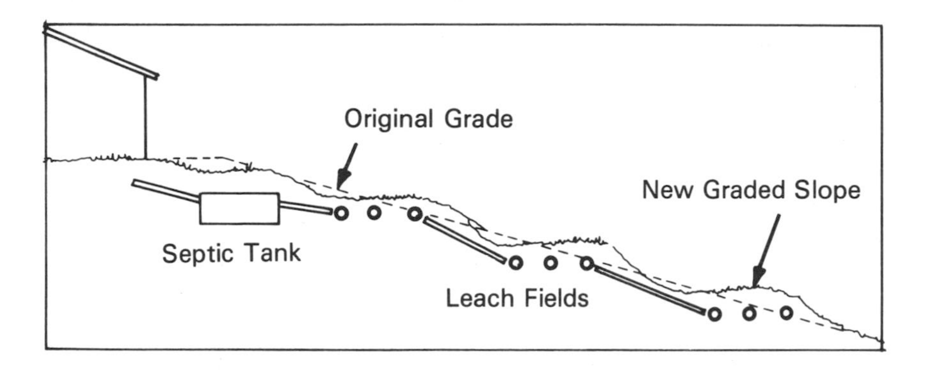

Figure 34. Sample Of Proper Leach Fields For A Sloping Lot

Distribution Box

The distribution box (see Figure 23, page 33) must be installed at a depth that places its outlet slightly above the leach field, so effluent will run from the box to the field. Solid pipes with watertight joints should connect the septic tank to the distribution system.

It's now time to determine the size of the septic tank you'll need. All sorts of figures are available as to the number of gallons a person uses per day, thus it's easy to figure the number flowing into the septic tank. Some figures are as high as eighty gallons per person per day. However, a more realistic figure is between thirty and fifty gallons.

Follow local codes in your area. If there are no regulations concerning the size of the tank, I recommend that any house up to four bedrooms in size should have at least a 1000-gallon tank. For each additional bedroom, increase the capacity by 250 gallons. This method of tank-size selection has proved to be more than adequate.

As with all guidelines, there are always exceptions. They include summer camps, roadside stands, gas stations, and other buildings. These should be handled on a case-by-case basis.

Digging the pit for the septic tank is one of the most important parts of installing a septic system. It should be a minimum of ten feet from the house, even though some codes say only five feet.

It must also be at the right depth. The septic tank's outlet pipe must be slightly above the inlet to the distribution box (one-fourth inch pitch per foot of run to the distribution box is usually about right). The pitch from the house to the inlet of the septic tank should be about the same. Too great a pitch leading into the tank, as I've said before, will mean sewage flowing into it will disturb the contents, resulting in solids flowing out of the tank and eventually blocking up the disposal field. If the pitch is too flat the pipe may become clogged.

And the bottom should be level. If it isn't, the care used by the manufacturer in making the inlet higher than the outlet is wasted.

Finally, the hole must be deep enough so the tank can be covered by about a foot of soil. As you'll remember, freezing is not a problem with septic tanks, so the tank doesn't have to be below the frost line. Also, sooner or later, the tank will have to be pumped out, and a deep covering of soil will only impede this work.

Remember one thing: tanks will float—even concrete tanks. Should you install a tank just before a sudden downpour, it's very apt to come right out of the ground. And it won't settle back after the rain into the right position or proper depth. The easy way to prevent this is to fill the tank with water after it's set in place. When water is not available during an installation, have a truck of water ready to fill the tank.

hold it down boy!

Then, somewhere along the way, you decide there will be a bathroom in the basement. The contractor doesn't care, and the plumber is delighted.

The sewer pipe now has to exit below the footings. This will put the tank eight to ten feet below grade unless you have a slope outside which will allow the tank to be a foot or so under ground, where it should be. However, to bury the tank at the proper depth may require putting it sixty to seventy feet away from the house—much too far.

At this point you'd better consider a "sewage ejector" or lift pump. These units pump sewage up and out to the tank. They are very reliable, and as they break up the solid matter before it reaches the tank, the chance of a stoppage is reduced and the bacterial digestion of the broken-up solids is greatly accelerated. In short, if plumbing fixtures are to be in the basement, by all means consider a lift pump.

Sometimes, because of the contour of the property, it's not possible to have a gravity-fed system to the leaching area. Years ago this would have been a problem. But today it is not. A second tank beyond the septic tank is required. It must have a submersible sewage pump. With it the effluent can be pumped through a one-and-a-half or two-inch pipe to the leaching area. This system allows a flexibility unavailable a few years ago.

When you install a tank, then bury it, be sure to measure and plot the position of the tank so that you can find it later. The best method is to measure from two points to the center cover of the tank. If we know where this cover is, the other two are easy to find. (Figure 35)

After you've taken the measurements, the best place to record them is inside the fuse box in the basement. Or use a crayon to draw a diagram with measurements on the basement wall. We used to write the measurements on the outside wall of the house, near the tank. But we'd return to find that painters had painted over the measurements. Love house painters.

When you're choosing a tank, and deciding between a steel or concrete tank, don't debate for long. The better selection is concrete. It's far stronger and often less expensive. Concrete tanks also seem to work better. What's more, I think the bituminous tar coating inside the steel tank tends to inhibit bacterial action. There are also liquid detergents on the market that, when used, can dissolve the tar on the inside of the tank.

You may want to build your own tank. There's nothing wrong with the homemade variety, provided you follow certain guidelines. I've seen every conceivable type of tank, from 55-gallon drums to a 3000-gallon oil truck buried in the ground. They've been made from logs, stone, blocks, wooden boxes—even railroad ties.

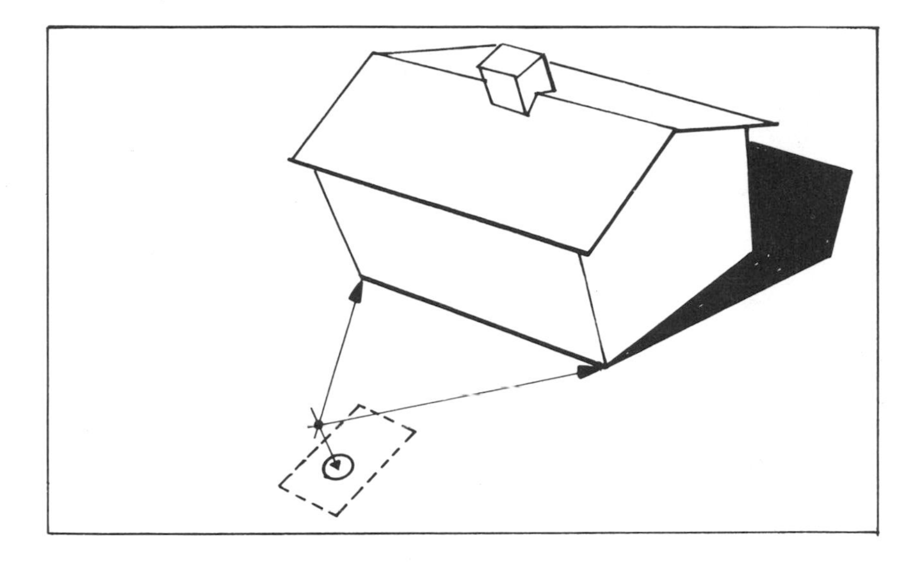

Figure 35. A Reference Sketch Of Septic Tank Clean Out Location For Your File

Do better than that. The best method is to use poured concrete walls and bottom. Next best is a tank made of concrete block, with the blocks mortared together. A box with the interior dimensions eight feet long, four feet wide and five feet deep, holds approximately 1000 gallons.

A reinforced concrete cover is best. Remember the cover will have to come off sooner or later, so make it with manholes, or make the cover out of several small slabs that can be handled by man with a bar. Don't forget steel loops for lifting. As second choice, a wood-plank cover can be used.

Placement of the baffles is most important. The outlet baffle is about six inches longer than the inlet baffle. A plumber can make these up for you.

CHAPTER 8

Commercial Tanks

The greatest problems we have today are with commercial tanks. By commercial tanks, I mean those connected to restaurants, hotels, motels, or other commercial establishments, which have kitchen wastes going into them.

In every case, the culprit is grease.

Grease can cause trouble throughout the system. It can plug up the inlet baffle so no liquid can flow into the tank, and gradually cause back up. Or it can work its way from the tank to the leach field, and there block up the spaces in the crushed stone, eventually spoiling the field.

Not long ago many such kitchens had grease traps, installed below the sink, but in front of the septic system. This trap would collect grease by cooling and trapping it. As it cooled, the grease changed from liquid to a semi-solid consistency much like butter. Occasionally the trap had to be opened and this mass of grease cleaned out and thrown away. Remember K.P. in the Army?

But that was in an era when pots, pans, and dishes were washed by hand—in lukewarm water with lye soap. The introduction of detergents and mechanical washers using much hotter water changed all that. Detergent puts grease into suspension in very hot water, causing it to slip past the grease trap and flow into the septic tank.

What can you do about it?

Fortunately, many things.

First, try to keep as much grease as possible out of the system. Cooks should learn to pour grease from a hot frying pan into an old coffee can. And kitchen workers should scrape dishes and pans carefully before putting them into the dishwasher.

Have the septic tank as close to the kitchen as possible. This will eliminate one of the most common problems we find. Grease-laden water gradually cools as it flows through pipes. If it travels too far, by the time it reaches the inlet baffle, the grease goes out of suspension and can build up in the inlet baffle, causing a blockage there.

Establish a schedule of inspections, to ultimately determine how frequently the tank should be pumped out. Regular checks will prevent grease from blocking up the leach field.

The best commercial system, in our experience, is a series of septic tanks in line. The effluent flows from one to the other before reaching the leach field. Any plumber or septic tank serviceman will tell you the amount of solids that get through the first tank is unbelievable.

more than one

If, for example, your restaurant requires a 5000-gallon tank, the best arrangement is to install a 3000-gallon tank followed by a 2000-gallon tank. And if the grease load is heavy, add still another 1000-gallon tank, to better protect the leach field.

You may find the first tank has to be pumped out twice a year, the second tank once a year, but the third tank only every other year. This can only be determined by regular inspections, until a schedule is established.

CHAPTER 9

The Free Advice Department

There's a shortage of many things in the world today, but free advice is not one of them.

The following little gems are a collection of general information that I either forgot earlier or simply feel the need to include now.

You've decided to buy a country house. The real estate salesman tells you all about the location, neighbors, schools, and stores, but rarely, if ever, is the water or sewage system ever discussed.

If you're like most people, you just assume that to have water, you just turn the little knob. To get rid of it, just push the little lever. In most cases, real estate salesmen know little about the water system and less about the sewage. Now is the time to call the local plumber and find out about the water system, what kind it is and what condition it's in.

Ask him, and anyone else that might know, about the sewage disposal system. What kind and size of tank? When was it last pumped out? Who pumped it? Then call the pumping company and ask questions. You may find out something about the condition of the leach field. Finally, where are the tank and the field? Don't hesitate to ask for this in writing.

Buying a house with a poor water or sewage system is like owning a sick cat. They never get better, only worse.

The question I'm asked most frequently is, "What can I put in my tank to make it work better?" These are the people who hope I'll tell them to drop in a yeast cake once every six months, and they'll never have any septic tank problems. The answer is always the same. The important thing is not so much what you put into a tank, as what you *don't* put in.

At the head of the list of things to avoid is colored toilet paper. It just does not dissolve, and I can only assume it's due to the dye impregnation of the paper. It is one of the major contributors to your septic tank problems, but very good for my business.

Next on the list are drain cleaners and toilet bowl cleaners. Most drain cleaners are basically sodium hydroxide, more commonly known as lye, which is one of the strongest caustics known. Bowl cleaners, on the other hand, are usually acidic in content. Both of these compounds will kill the bacteria in the tank, and without these little microbes, nothing will decompose. Many commercial cleaning products state, "Will not harm septic tank." This is true, but they sure raise hell with the contents. With this in mind, use them sparingly.

Lye can cause a major problem, and you'll understand why if you understand how lye works with grease. When there's a stoppage in a pipe because of grease, lye in the form of drain cleaner is added to the water. Heat is generated, and the stoppage, softened by the heat, clears itself. Fine.

But what happens next? Some of you may remember how your grandmother made soap. She took the grease she had saved, put it in a pan on top of the stove, and heated it. Then she added wood ash, or caustic soda, stirred the mixture, allowed it to cool, and cut it into squares. She called it lye soap.

Basically, the same thing happens when lye is added to a grease stoppage. The trouble shows up as the "soap" blocks the inlet baffle of the tank, since the grease has cooled and begun to harden by the time it reaches that baffle. The only way I've found to avoid this type of blockage is to run hot water into the sink regularly for three or four minutes, once or twice a week to clear the grease, instead of adding lye. If you have an automatic dishwasher that empties into the same pipe as the one that drains the sink, this will work even better, as it runs the hot water for you.

Next on the list of things to avoid is bleach. As you know, bleach is nothing more than a diluted chlorine solution. Chlorine is a disinfectant and bactericide. It's used in swimming pools and sewage treatment plants for water purification. Use it sparingly. It will kill the bacteria in your tank.

Finally, coffee grounds. I don't know why, but tanks that have large amounts of coffee grounds dumped into them work poorly. Put those grounds into your compost pile or trash.

The question next most frequently asked, is, "How often should I have my tank pumped?"

The old rule of thumb was to pump a 300-gallon tank every three years, and a 500-gallon tank every five years. This rule no longer hold true.

Today no two tanks work the same. One system will fail before another, in cases where conditions are nearly identical. Summer camps and vacation homes have systems that seem to go on indefinitely between pumpings. Some homes need the tanks pumped every year.

As a general rule, if you have a 1,000-gallon tank, and four persons live in the home, plan to have the tank pumped every five years. Ask the person who does the pumping if you've waited too long, or if you could have waited another year. That way you get a schedule started.

You should know the location and size of the tank in your system, as well as when it was last pumped out. Knowing this will help you to sleep better at night. Should you have doubts, call your plumber and ask his advice. He'll probably recommend an inspection. Once this is done you have a starting point, and can establish a schedule.

What about additives? On the market today are any number of products their manufacturers claim do all sorts of good things for your tank. These benefits include everything from eliminating the need to ever dig up your yard to pump the tank, to rejuvenating a tired, old disposal system. The septic tank's Geritol.

In all fairness, it's hard to judge the merits of these products. I'll explain why. People are strange creatures. "Out of sight, out of mind" applies to their septic systems. It's only when the tank backs up that they add these chemicals, and, of course, by then it's too late. The additions don't work. Since Joe Public never reads the directions anyway, and seldom follows them when he does, this is the result.

Years ago I sold an activator for septic tanks, and left specific instructions on how to use it. You know, a little bit in the tank every now and then. Several years later when we returned to service these tanks, sure enough, there in the garage the product sat, never even opened. So I quit selling it. Just pumped out the tank and kept my mouth shut.

I've put liquid enzymes into many tanks after pumping, in an attempt to help the bacteria start the digestive process. But this is the only time I advocate adding anything.

Experience shows that the color of the liquid in a tank is very important in evaluating the bacteria action. Colors range from a pale yellow, to various shades of grey, and finally black. The darker the liquid, the better the tank is working bacteriologically, and when we pump these tanks with dark contents, no additives are needed. They should be used only when the lighter colors appear.

Every day I hear people say they have a garbage disposal unit, but don't use it. The excuses are endless. Rather than attempt to list them all, let me say there is no evidence to substantiate any reason for not using a garbage disposal.

I have had the opportunity, over twenty-five years, to monitor many tanks, both with and without a grinder in use, and I can honestly say I think these tanks work *better* with the addition of ground garbage. This, of course, doesn't include the silverware my wife manages to feed ours. My advice is: if you have a kitchen pig, use it. If you don't have one, get one. I can think of no cheaper way to have your garbage collected than to have it pumped out of your septic tank every five years or so, along with everything else.

Every now and then someone asks me if it's true coffee grounds will keep the sink pipe clean. The answer is no. In fact, nothing will clog a pipe quicker than coffee grounds and grease.

If you have an old vertical steel tank and must remove the entire cover to pump the tank, there's probably an easier way.

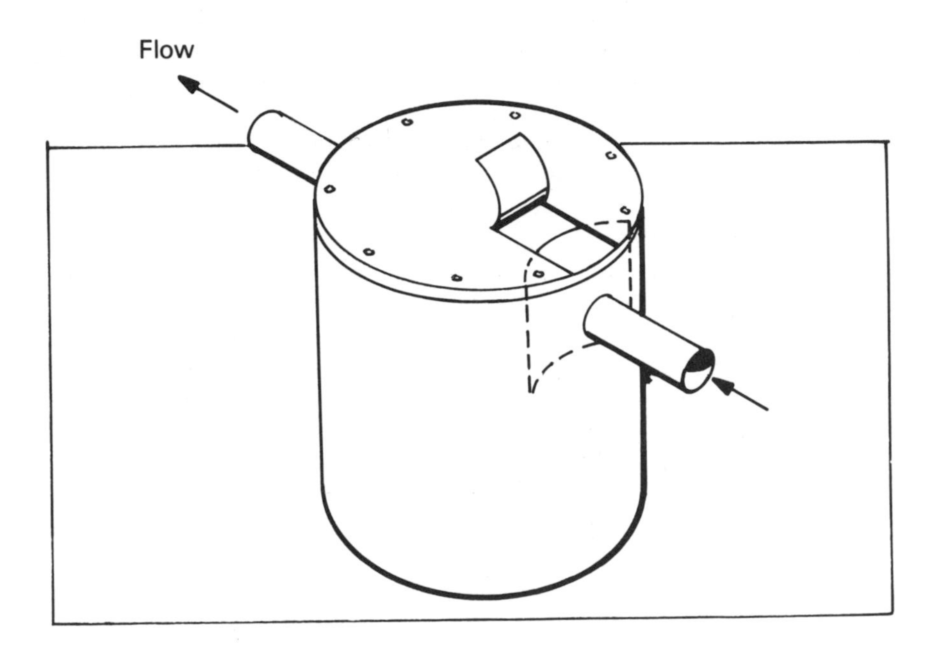

Figure 36. A Septic Tank Cover Cut For Inspection, Rodding Or Pumping

Dig down to the cover at the edge nearest the house. Then find the inlet pipe. It will be about two inches down the side of the tank. Using a three-pound hammer and a sharp cold chisel, cut the cover above the inlet pipe as shown in the illustration. When you lift this cut section, you'll see why we cut it where we did. From this opening over the inlet baffle, we can inspect it. We can rod to the house and pump the tank properly if necessary.

Whenever you have the tank pumped, flush the toilet before replacing the cover to make sure the line is open. A wad of flushed toilet paper will help determine whether a partial blockage remains. Now is the time to find out, when the tank is open. It could save more digging later.

If you have cut into a steel tank, as suggested above, fold the cover back in its original place and lay a piece of tar paper or plastic bag over the cut sections. (Figure 36)

If for some reason the entire cover has to be removed, mark the cover in some way so that it can be put back in the same position. This is particularly important if the cover has clips, since these covers are not exactly round, and replacing them can take much time and effort. A little scratch on the cover eliminates all this.

Keep These Things Out of Your Septic Tank

1. Grease
2. Colored toilet paper
3. Drain cleaners (lye)
4. Toilet bowl cleaners
5. Bleach
6. Coffee grounds
7. Chemicals (such as those used in photography) that will kill bacteria.

Keep These Things in Your Garage

1. Sink plunger
2. Toilet plunger
3. Toilet auger
4. Small sewer snake
5. 14″ pipe wrench

As long as we're talking about covers, I'll tell you how to get the ones off concrete tanks. They can be difficult to remove, particularly the first time. First put a bar, chain, or rope through the little steel loop in the cover. As you pull up on this, give the edges of the cover several sharp wallops with a heavy hammer. The cover will pop up. Otherwise, you could be there all day.

Pumping isn't always the answer to every problem. The answer may be to clean out a simple stoppage in the inlet baffle, a plug in the inlet pipe, or a problem in the outlet area.

When you have the tank open, the accumulation of solids in the digestion area should be checked and the depth estimated. If you find the thickness of the floating material is only three inches or less, all you need to do is to break up this matter with an old hoe or rake. By doing so, you're putting the dried floating cake back into the liquid where the bacteria can again digest it. This can extend the length of time between pumpings considerably.

But if the thickness of this floating mat is more than three or four inches, the best thing to do is pump it out.

Quite often we find that a tank has been placed too deep in the ground. This means that when the tank has to be opened, much of the lawn in the area has to be dug up.

In these cases we generally put a fifty-five-gallon drum over the pumping access hole. In the case of kitchen tanks, we put one over the inlet baffle where grease accumulates and most stoppages occur. Also, it makes for easier rodding should the inlet pipe become clogged. It's also wise to check the inlet area after pumping a tank.

We agree with what you're thinking. This shouldn't be necessary, and tanks shouldn't be placed that deep in the ground. But some are, and placing a drum over the strategic areas is much easier than digging so far each time something goes wrong.

The best drums for this are called lacquer barrels. They have removable covers. If these are not available, use any fifty-five-gallon drum. But nothing smaller. Many times a plumber will run a four-inch pipe to the surface with a cleanout cap on top. This system will not work well because it's *impossible* to pump a tank properly through a small pipe.

If the tank is way down, don't use one drum on top of another. If the depth is over three feet, use three-foot or even four-foot well tiles.

To use a barrel over a tank, a little preparation is necessary. If you can find a barrel with a removable lid, it's necessary to cut the bottom out with a hammer and cold chisel before placing the barrel over the pumping cover or opening. If all you can find is an old oil drum, that will do nicely. Again, cut out the bottom, but save the round portion you've removed. Now turn the barrel upside down and cut the other end as pictured here.

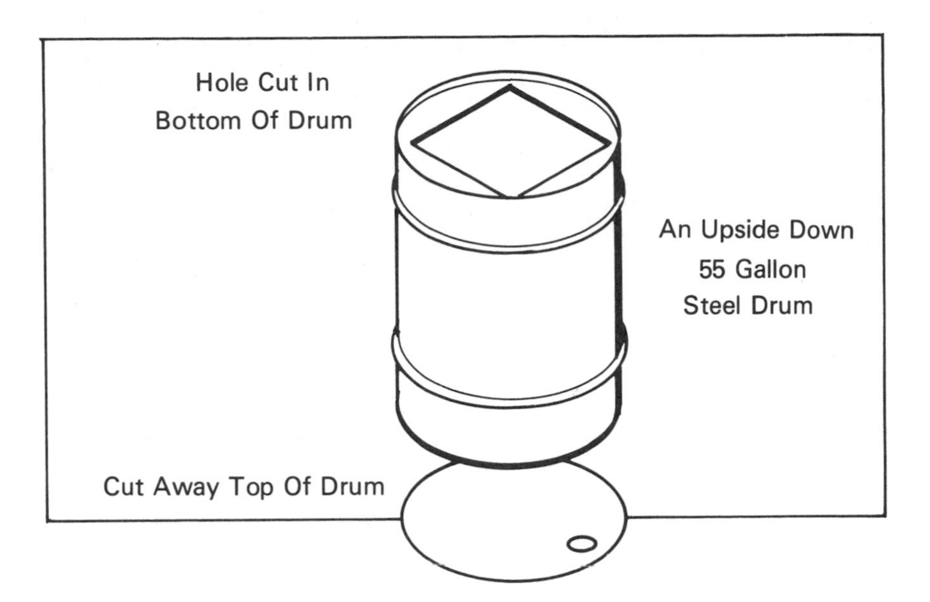

Figure 37. A Suggested Method For Inspection Cover Extension

Place the barrel over the tank opening, with the cut-away end on the bottom. Place that round cut-away section on top of the barrel. The next time you need access to the tank, dig down to the barrel, remove the top cover. And there you are.

Over the years we've installed hundreds of these barrel extensions. They have saved us lots of time and labor, and have saved the customers a lot of money.

Should the drum be too tall for the hole, use the cold chisel to cut it down to proper size.

Sometimes it's necessary to conserve water. The leach field becomes saturated, or the spring or well threatens to run dry. There are many ways to save water.

The greatest consumption of water is in bathing and washing clothes. Next come the toilet and lavatory, and finally, food preparation.

Try to do larger and fewer wash loads. Next, buy a water-saving shower head. It will use two and one-half gallons per minute, compared to the six or seven that flow through a conventional head. If you have teenage daughters, this is a must. With smaller children, it isn't necessary to fill the tub to the top, and teddy bears don't need a bath every week.

Some people put bricks and other objects in toilet tanks to reduce the per-flush flow. An easier way is to bend the float arm down to lower the water level. All that's required is enough water to clear the material in the bowl.

at least every day

In rural areas, people are always concerned about the quality of their water. There are two tests for drinking water. One is a bacteria or "coliform" analysis, which will tell if your water is contaminated. Coliform bacteria are intestinal bacteria found in all mammals. Although these bacteria are not harmful to man, their presence in water indicates the possible presence of disease-carrying bacteria. If you have any questions concerning the quality of your water, call your local health officer and get an explanation of the procedure for having your water tested.

The other test is called a "chemical" test. This is taken when there is a taste or an odor in water. The water may also be turbid or cloudy. Plumbers will send samples of such water to laboratories or water-treatment equipment manufacturers for an analysis. They will reply with a complete report and recommendations for solving the problem. There is usually no charge for this service.

If you have a spring or shallow well (twenty-five feet or less in depth), you may want to chlorinate the water occasionally. First, estimate the capacity of the system. Then, for every 500 gallons of water, add one quart of Clorox to the spring or well. Let each faucet in the house run until you can smell the chlorine.

The best time to do this is the last thing at night, so the chlorine can work for several hours in both the spring and the pipes. You may want to put away a little water in the refrigerator before you chlorinate, for your coffee in the morning. The chlorinated water is not going to taste too great for a few days.

This chlorination should be done twice a year for best results. A good time to test the water is after one day of normal use. It's good to get advice on chlorinating from your local water department, health officer, or the state health department. Do not use swimming pool disinfectants, only chemicals specified for water purification.

If you have a deep well, the chances are pretty remote that the water is contaminated. But if there's the slightest doubt, get it tested, even if it's just for peace of mind.

The only problem I haven't been able to solve is how to get my wife to buy white toilet paper. Since we have a blue bathroom, it goes without saying that we must have blue toilet paper.

One day I laid down the law and declared, "Alice, from now on you are going to buy white toilet paper!"

Sure enough, the next day she returned with four rolls of white paper. Complete with blue flowers.

I guess I'll have to paint the bathroom white.

CHAPTER 10

Summary

In conclusion, I'd like to say that the work I do has never been boring or dull. And since I started alone, I've had the opportunity to make all the mistakes possible—and without any outside help. It has also given me the opportunity to meet all kinds of people, rural Americans. There are no finer people in all the world.

There is one thing, however, that puts this business in a class by itself, and that's the fact that no one ever tells you how to do it. No one even volunteers to help. They're not interested in how you do it or what you do it with. Just get it done. And quickly.

I sincerely hope that somewhere, someone will get something useful out of *Country Plumbing*. After all, that was the intent of this book.

—so long folks!